Contemporary Issues in
Biomedical Ethics

T0321724

Contemporary Issues in
Biomedical Ethics

Edited by

John W. Davis
Barry Hoffmaster
and
Sarah Shorten
The University of Western Ontario
London, Ontario, Canada

The HUMANA Press Inc. • Clifton, New Jersey

Library of Congress Catalog Card No.: 78–71406

Davis, John, Hoffmaster, Barry, and Shorten, Sarah
  Contemporary Issues in Biomedical Ethics.
Clifton, N.J.: Humana Press, The
320 p.
7902   781107

The present volume is based on and augmented from papers delivered at the Colloquium on Biomedical Ethics held at the University of Western Ontario, London, Ontario, Canada, October 27–30, 1977.

Printed in the United States of America

FOREWORD

Not long ago, a colleague chided me for using the term "the biological revolution." Like many others, I have employed it as an umbrella term to refer to the seemingly vast, rapidly-moving, and frequently bewildering developments of contemporary biomedicine: psychosurgery, genetic counseling and engineering, artificial heart–lung machines, organ transplants—and on and on. The real "biological revolution," he pointed out, began back in the nineteenth century in Europe. For it was then that death rates and infant mortality began to decline, the germ theory of disease was firmly established, Darwin took his famous trip on the Beagle, and Gregor Mendel stumbled on to some fundamental principles of heredity.

My friend, I think, was both right and wrong. The biological revolution did have its roots in the nineteenth century; that is when it first began to unfold. Yet, like many intellectual and scientific upheavals, its force was not felt for decades. Indeed, it seems fair to say that it was not until after the Second World War that the full force of the earlier discoveries in biology and medicine began to have a major impact, an impact that was all the more heightened by the rapid biomedical developments after the war. The emergence during the war of antibiotics, and later of effective pesticides, reduced death rates still further in developed nations and had a startlingly powerful effect in the developing nations—so rapidly did death rates decline in those nations that a new problem was born, that of excessively high rates of population growth. That was indeed a biological revolution for those nations. Rapid developments in surgery, in chemotherapy, in contraception, and molecular biology were all, taken together, further signs that the impact of biomedicine on human life was going to be profound and long-lasting. In that sense, then, we are only now beginning to see the full scope of the changes that may be in store for humankind.

The first responses to these developments were astonishment and enthusiasm. For the first time in human history, it began to be possible to dream of a conquest of some of our oldest enemies—early and untimely death, the birth of defective children, organically determined violence, the vagaries of a body all too prone to sicken and die. Here was "progress" in the raw, and now a form of progress that held out the tantalizing hope that the miracles science had wrought in the external world of nature could be duplicated in the inner world of the body and mind. In that light, it was hardly surprising that the biomedical sciences gained massive public support and funds, and that biologists and physicians gained a special prominence in society.

By now, however, the enthusiasm has cooled somewhat. Most of the early and dramatic "conquests" of medicine were in the area of infectious disease: smallpox, scarlet fever, whooping cough, and polio. The development of potent antibiotics represented still another "miracle," with startling effects upon mortality and morbidity rates. The language of "conquest" and "miracle" is rarely heard these days, and with good reason. A whole range of chronic conditions, among them cancer and heart disease, have so far proved particularly resistant to dramatic breakthroughs and instant cures. Clearly, with respect to those conditions, we are in for a long period of further research and slow, halting progress.

No less important, it is also turning out that the price of success is becoming higher and higher. It is now possible to extend the lives of some individuals in ways not even imaginable a few years ago—the artificial respirator is a convenient symbol of that power. Radical surgery, chemotherapy, and a variety of other medical techniques and technologies have made a difference. But they have also created some terrible dilemmas. For the power to extend life, or to relieve physical suffering, has not been matched by an equal power to create happier or more satisfying lives. This is most evident with the growing proportion of the aged in our society. It is primarily better nutrition, sanitation, and medical care that have created such large numbers of the elderly. Yet we know that life for many of those who have been spared the ravages of childhood disease, or death by sudden infection, can simply be a misery—a misery because of the continuation of unconquered chronic disease, or a misery because of a social structure that simply does not know what to do with a large proportion of old people.

The very pervasiveness of that problem often causes it to be overlooked altogether when issues of medical ethics are discussed. It is the more dramatic dilemmas that catch the eye: a Karen Ann Quinlan case, or an instance of sterilization abuse, or the advent of amniocentesis and prenatal diagnosis. Those surely embody important problems, but they should be seen within a context of a general extension of individual life-spans, where they may seem less wrenching though their extent is simply enormous.

Nonetheless, within that context, one seems to see nothing but instances of the mixed blessings of advanced technological medicine. Indeed, it is safe to offer one solid proposition about such medicine: every step we take toward "progress" will be accompanied from now on by a serious moral dilemma. The dilemma may be wholly moral in character, as is now the case for those parents who must decide whether they want to abort a defective fetus prenatally diagnosed to have a serious genetic defect, or for those parents who must decide whether to authorize a life-saving operation for a damaged newborn when the prospect of salvaging no more than a permanently and severely crippled child is great.

In other cases, the dilemma will embody a combination of moral and technical ingredients. The debate over recombinant DNA research has been both a technical debate—what are the real potentialities for good from that research, and the real potentialities for harm?—and a moral debate—do we have the right to pursue great gains at the possible price of great losses? Much more common of course are those daily moral dramas in which a patient or a family must decide whether to risk a hazardous procedure that may save life, but also may produce, for the life thus saved, an existence of prolonged pain and suffering.

Dilemmas of this kind are forcing some fundamental reevaluations. First, they are forcing us to ask what exactly we should hope for from medicine. Should its goal be to forestall death, to relieve suffering, or simply to maintain health? Those are different types of goals, and although they may overlap, a choice of one of them will have massive consequences for establishing research priorities and organizing national systems of health care delivery. Second, the dilemmas are forcing us to reexamine some traditional moral values. Does the principle of the "dignity of life" or "the right to life" require that medicine always strive to keep people alive to the bitter end? But if we decide it does not, then what criteria will we use in deciding when to cease interfering with the process of death? Very strict and bureaucratic regulation of human experimentation can slow down clinical research, possibly at the cost of some deaths when a cure's development is delayed. But the notion of using human beings as guinea pigs, without their informed consent, is hardly less abhorrent.

The third occasion for reevaluation is the changing relationship between the doctor and the patient, and between the systems that deliver health care and those receiving such care. Quite apart from a general trend against the paternalism of experts, physicians included, the new moral dilemmas in medicine force the physician and the patient into a new relationship. Decisions must be made that go well beyond mere scientific knowledge, into areas that are clearly moral and clearly bear upon trans-scientific ideas of the value of human life. Decisions of that kind can often only be made by the patient, or by those entitled to act in the patient's behalf. That some physicians, and indeed some patients as well, resist this insight does not negate its truth. Analogous problems appear at the level of decisions about basic research and the delivery of health care. The decisions to be made are political and moral, however heavy the technical components; thus is being born a new relationship between medicine and the public.

This book touches upon most of these problems. It would be nice to believe that, with enough thought, they can easily be worked through. That is most unlikely. Progress in moral thought is possible, however slow. Yet what is unique about our present situation is that as one problem is solved, the pace of medical advance guarantees that a new one will take its place. And often enough, the direct price to be paid for the solution of one problem is the immediate creation of another. That means we are in for a long period of moral analysis and anguish. There is nothing here that can be solved once and for all. Instead, the great virtue we now need to develop is that of being prepared to wrestle with the problems over a long period of time, to accept a slow movement, and to build into our very way of looking at the issues a moral perspective which is sophisticated and rich. That will not be easy, but the present book is surely a strong indication that the challenge is not being evaded.

Daniel Callahan

The Hastings Center
Institute of Society, Ethics and the Life Sciences

PREFACE

Recent theoretical and technical advances in biology and the health sciences have precipitated extensive reassessments of both medical practice and societal values. These changes are occurring at a pace that renders awareness of contemporary debates in these areas the urgent responsibility of theoreticians and practitioners alike. The papers and commentaries collected in this volume were delivered at a Colloquium on Biomedical Ethics at the University of Western Ontario in October, 1977. The colloquium was sponsored by the Faculty of Medicine and the Department of Philosophy, in conjunction with the Faculty of Law, and indeed this cooperation is an indication of the mutually felt need of scholars in diverse disciplines jointly to confront such problems at this time. The interdisciplinary nature of the volume reflects the belief that the complementary knowledge and skills of lawyers, philosophers, and a variety of health-care professionals can, and indeed should, profitably be pooled in the effort to understand the complex dimensions of bioethical problems.

Selected bioethical issues are discussed in this book. It is intended as a suitable anthology for general reading in the field, as well as a reasonably balanced and diversely representative volume suitable for classroom use and discussion. The articles remain throughout at a technical level accessible to the nonspecialist or the undergraduate, and the multidisciplinary approach makes it particularly appropriate for use in medical schools, as well as in law or philosophy programs. Several contributors have expressed their hope that the present promise of continued interaction between the professions will be fulfilled, and will be conducive to both the enlargement of all our knowledge and to an increased understanding between health care professionals and society as a whole.

The papers are organized on the basis of content. The titles of the sections capture the themes shared by papers in each. All of the papers, however, fall into two general categories. Some deal primarily with metaethical issues, that is, with how moral decisions in biomedical ethics should be made. Others address particular moral problems that arise in biomedical contexts.

John Ladd's essay, "Legalism and Medical Ethics," is a good introduction to the manner in which metaethical issues are generated by problems in biomedical ethics. Ladd argues that moral problems in biomedical ethics should not be discussed in the language of rights. He believes that the concept of a right is appropriate in the legal context, where it has its home, but is not appropriate for the handling of the many moral problems that arise in medical ethics. An appeal to one's moral rights is proper when a conflict is impersonal, for example, in a clash of interests between strangers, or between an individual and a formal institution such as a hospital. But in situations in which personal considerations are relevant, an alternative moral approach is needed. Thus, for example, in the personal relationship between a patient and physician, an appeal to one's rights is often inappropriate, improper, or even immoral. As an alternative Ladd recommends an ethics of responsibility, an ethics of giving and receiving, founded on the principle of matching what the physician has to give with what the patient needs. Both parties should strive to foster the moral integrity of their relationship, rather than view the relationship as one involving mutual rights and obligations. One consequence of Ladd's view is that a physician's decision about the "responsible" action in a situation may conflict with what a patient has a right to. But acting responsibly on the part of a physician requires that what promotes a patient's welfare always overrides what respects a patient's rights. So, as Ladd says, "Simply to give in to a person who refuses something needed, say, to stay alive may, in some cases, amount to an abdication of moral responsibility."

Robert Audi's paper, "The Moral Rights of the Terminally Ill," proves a nice counterpoint to Ladd's discussion of rights. Audi is concerned with the resolution of moral problems that arise with respect to terminally ill patients, but his arguments commit him to a metaethical position opposed to Ladd's. Audi claims that the main source of problems in medical ethics is the conflict of rights. He is prepared to recognize a large number of moral rights because of his beliefs that infinitive grammatical constructions should be used to express rights and that a unique right should be countenanced for each distinct state of affairs expressed by an infinitive construction in a true attribution of a right to someone. Thus since it is, on Audi's view, true to attribute the right to refuse medical treatment to everyone, there is a distinct right to refuse medical treatment. At the same time Audi realizes that the concept of a right requires clarification. His essay is an attempt to help one understand what a right is and how conflicts between rights can be handled. Audi begins by arguing that there is a morally significant distinction between killing and allowing to die. In an examination of the moral rights of the patient, physician, and family, he discusses when terminal patients should be allowed to refuse or terminate medical treatment, and what terminal patients should be told about their condition. Audi accepts such rights as a right to control one's body, a right to end one's life, and a right not to be lied to about one's condition, although he holds that these rights, and the others he recognizes, are not absolute and can be overridden by more important moral considerations. Audi's essay concludes with a brief discussion of the legal implications of his moral view.

Later papers by Dyck, Flaherty, and Roy also ask what considerations are relevant to moral decision-making in biomedical contexts. Arthur Dyck's paper, "Professional Education for Humane Care," deals with the obligatory nature of compassion in both medicine and morals; the ancient question of the teachability of virtue is raised. A common complaint about medical education is that although medical students are taught technical skills, they are not taught to be kind and compassionate, and in fact may even lose their compassion during medical school training. For Dyck the attitude of compassion is morally obligatory, but compassion is more than an attitude. Compassion is a moral virtue, involving the ability to discern what is right and wrong and the willingness to do what is right. In addition to being a necessary component of morality, compassion is necessary to the successful practice of medicine. It promotes the injunction "first of all to do no harm," motivates physicians to inflict pain, suffering, and risk of death in order to remove greater pain, suffering, and risk of death, enables physicians to diagnose illness properly, and finally encourages the comfort and hope that are essential to therapy. Dyck rejects arguments for the claim that compassion cannot be taught and defends his view against the objection that irrational emotions and attitudes should not enter the process of moral decision-making.

In her paper, "Accountability in Health Care Practice: Ethical Implications for Nurses," M. Josephine Flaherty emphasizes the independence that nurses have because they are professionals. Being a professional means that nurses must be prepared both to exercise judgment in making decisions and to be accountable for such decisions. In fact the most important characteristic of a profession, Flaherty argues, is its accountability, the taking of responsibility for one's own professional actions. Nurses, as independent health care professionals, are accountable *for* their behavior rather than merely accountable *to* someone. One of the most important responsibilities of a nurse is to preserve the safety, dignity, and autonomy of a patient. The relationship between a health care professional and a patient should be used to enhance the autonomy of a patient. Central to the process of caring is encouraging patient autonomy; the goals of caring should be the growth of the person to whom care is given and the actualizing of the person who gives the care. Accountability means, moreover, that nurses must adopt a critical attitude toward the actions of other members of the health care profession, the situations in which they find themselves, the customs and taboos that serve as norms in the contexts in which they practice, scientific and technical research, and any biases or prejudices that may determine their own behavior. Clear answers to moral problems will not be provided by simply questioning established policies, institutions, and ways of acting, but such a critical attitude will provide an appropriate atmosphere in which a professional can address such problems.

In "The Public Responsibility of Philosophy" David Roy argues that the community of philosophers has a public responsibility to develop a system of norms for regulating the immense power of the scientific–technological community. The task is to develop an objective global ethic that will, since it must deal with the power to affect the foundations and structure of human nature, consist of norms based on judgments about what is worthwhile and essential. This task is a distinctly philosophical activity, and the philosophical community must accept the challenge on behalf of the total human community. After examples to show how recent and prospective scientific and technological developments in biomedicine raise philosophical questions, Roy sketches the foundation of a global ethic for handling these questions. Any attempt to develop such an ethic must confront a dilemma:

"But how can philosophy hope to discover these foundations for an ethics capable of guiding our use and development of science and technology if the very source of the principles for such a foundation—the nature of man and the nature of things—are subjected to radical change by the very activities we seek to control?" Roy argues that an analysis can be provided that yields a rational, objective, and therefore global ethic founded on a principle discoverable in the nature of things. The possibility of argumentation in logic, science, and scientific technology presupposes shared values. From such values emerge two specific ethical principles that Roy claims are central to the development of a global ethic. This kind of argument is transcendental. In summary Roy says, "the community of communication as the presupposition of all rational discourse constitutively demands an ethic of survival for the sake of liberation," or, in other words, a global ethic.

Openness of communication is also a central theme of Lisa Newton's paper, "Hippocrates Lost, A Professional Ethic Regained". She argues that the very existence of popular and public debate on bioethical issues is an index of profound and important changes in the nature of the medical profession, changes that she hails as progressive and morally significant. From a profession that protected its members, its traditions, and its authority behind a curtain of secrecy, medicine has evolved from the age of Hippocrates to its present status, wherein it acknowledges its openness to rational argument and criticism, and the public's right to share its knowledge and enter into its ethical dilemmas. In the face of its present need for an ethic to replace the obsolete ethic of Hippocrates, Newton argues that the medical profession must acknowledge the need for society as a whole to debate its priorities in order to define adequately the "health" which is the guiding light of medicine. Since the right to practice medicine rests on possession of knowledge, research is an integral part of the practice of the profession as a whole, and the new ethic must take into account this fact. It must place value on the acquisition of new knowledge, and the progress of the profession. And it must take cognizance of the fact that patients have the right and the duty to understand their problems, the alternatives, and their consequences, and to take part in the evaluative decisions that must be made in the course of treatment. By protecting the privacy of the patient, and controlling the "paternalistic" autocracy of health-care professionals, the medical profession must proclaim its respect for the dignity and the autonomy of those in its care. Thus she urges the adoption of the contractual model for interpreting the doctor-patient relationship, and looks to the formulation of a "corporate" ethic that will provide a unified direction amid the diversities of the profession.

Michael Bayles would like to see the contractual model of the doctor–patient relationship replaced by a fiduciary model, analogous to that of the relationship between automobile owners and the mechanics hired to work on their cars. He argues in "The Physician as Body Mechanic" that the relationship between a primary care physician and a patient with a nonfatal physical disease is similar to that between mechanic and owner in that the professional must be licensed to "practice," and must obtain explicit authorization to perform any more than minor "repairs." Knowing more than patients, physicians must bear the responsibility of the trust that is placed in their competence and reliability. In both relationships, there is room for the concepts of both "cure" and "preventive maintenance"; what "should" be done is, in part, a function of the "owner's" values and the "mechanic's" public responsibility, and although the owner knows less, the right to fully informed consent to any procedure performed by the professional cannot be abridged. These considerations lead Bayles to the conclusion that the fiduciary model is more appropriate than any of the three alternatives he mentions: the paternalistic model fails to take account of the autonomy of the patient; the agency model fails to accommodate the superior knowledge of the doctor; and the contractual model does not capture the vulnerability of the patient and the consequent need to rely on the expertise and openness of the doctor.

The papers of Dickens, Brock, and Graber also discuss the location of decision-making responsibility in the physician–patient relationship. The requirement of informed consent presupposes that the patient is in many circumstances the appropriate decision-maker. In cases where paternalism seems justified, however, and involuntary civil commitment is merely one example, other people appear to be morally entitled to make decisions on behalf of the patient.

Bernard Dicken's paper, "The Ethical Content of Legally Informed Consent," is a discussion of the legal and moral dimensions of informed consent. He points out that legally informed consent is a matter of substance and not form. In other words, a signature on an informed consent document is not conclusive evidence of informed consent—this "apparent" consent can be shown, in a number of ways, not to be truly "informed." Dickens sets out the information that must be communicated to a patient in order for the patient's consent to be truly informed, as well as the information that need not be disclosed. He then discusses the ethical function of informed consent. Informed consent is a way of enhancing the autonomy of a patient. Allowing a patient to be a partner in making a decision about treatment protects the patient's intellectual dignity and bodily integrity, and thus protects the patient's status as a human being. Dickens goes on to raise a number of moral problems with informed consent. For example, if it is the patient who makes the decisive choice about treatment, thereby becoming morally and legally responsible for the consequences of the treatment, must a physician tell a patient that the patient consents at his own risk? And what does a physician do with a patient who does not want to be informed? Can a physician morally force a patient to be autonomous?

Dan Brock's paper, "Involuntary Civil Commitment: The Moral

Issues," deals with the moral justification of the use of coercion toward the mentally ill: when is it morally permissible to treat mentally ill persons against their will? There are two common justifications for involuntary civil commitment: patients are dangerous to others or to themselves. Brock analyzes the former justification by comparing it with the moral arguments used to defend criminal punishment. There are two striking differences between the approaches to civil and penal commitment. One is that society is more willing to confine wrongfully the mentally ill than it is the innocent. The other is that society is willing to accept preventive detention as a standard procedure in handling the mentally ill, but not in handling criminals or potential criminals. Brock examines several ways in which these differences might be justified. The most promising rests on a view of law as a choosing system; the law gives everyone a choice between forbearing from some action or being punished—"do it or else." The mentally ill are not capable of making this choice and so must be subject to an alternative analysis on the basis of likely costs and benefits. Mentally ill persons are committed involuntarily if the benefits of doing so promise to exceed the harms. Brock argues that this approach involves not treating the mentally ill as persons. Moreover, it is rarely, if ever, possible to show on empirical grounds that the benefits exceed the harms. Justifying involuntary civil commitment on the basis of dangerousness to oneself raises the issue of paternalism. Brock argues that we would want others to act paternalistically toward us only when we suffer defects of reason that impair our ability to form purposes, weigh alternatives, and act on the result of this deliberation. The troubling moral problem here lies in distinguishing desires or weighings that are irrational from those that are merely unusual. Brock's conclusion is that the common view, which holds that paternalistic civil commitment is more morally suspect than civil commitment on grounds of dangerousness to others, is wrong. Although there are serious moral problems with either justification for involuntary civil commitment, paternalistic civil commitment is less morally problematic.

Glenn Graber, in "On Paternalism and Health Care," questions the appropriateness of paternalistic action by health care professionals in any context. Graber defines "paternalism" as action taken on behalf of others without their consent and alleged to be for these others' good. He argues that its justification requires fulfillment of three conditions: knowledge of what is in the best interests of another, belief that the other person is unlikely to adopt the requisite course of action, and a position of authority to act in the other's behalf without specific consent. He points out that doctors, to the contrary, are authorized to treat a patient only when a professional relationship has been explicitly established between them, and then only minor procedures are permitted in the absence of additional authorization, which must always be made voluntarily and in the light of complete information. As a consequence, in this system, he argues, there is no room for paternalism as defined above. Nor should there be. Decisions regarding alternative modes of treatment, according to Graber, often require value judgments regarding the relative merits of each. These value judgments ought to reflect the values of the patient affected, and not those of the doctor, whose expertise is in medicine, not in ethics. Moreover, even if the values of doctor and patient should coincide, it is of critical importance that it be the patient who decides; only thus can the fundamental principle of preserving autonomy in ourselves and others be upheld.

The rights of patients are also explored by Joseph Ellin in "Sterilization, Privacy, and the Value of Reproduction". He concludes that amid the various irrational preferences some might have for procreation of one's "own" children over adoption, there are some preferences that are rational. But even these, Ellin argues, are not "important." Among these putatively rational preferences, he counts the value placed on pregnancy as an experience in itself, the importance that might be attached to having children with intelligence or facial features like one's own, and the "spiritual values" connected with the hope and expectation of begetting and bearing a child. Ellin contends that none of these is a preference with sufficient importance to ground any claim to have the "right to make up one's own mind" whether to procreate. None touches intimately the self-concept of the individual, nor has strong ethical, religious, or political import. None is connected to the right of self-expression nor the right to private, harmless pleasure. Procreation is not a matter in which the individual is clearly in the best position to make the decision. In spite of this, involuntary sterilization does violate, in Ellin's view, another important facet of the individual right to privacy: namely, the right to the use of one's own body, insofar as the sterilization procedure, if involuntary, invades one's body. Therefore, although he argues that it is incoherent to claim that sterilization violates the rights of the unborn, involuntary sterilization does constitute violation of a right.

Robert Baker shares with Ellin doubts about the intelligibility of statements some ethicists make regarding the unborn. In his paper, "Protecting the Unconceived," he examines, in particular, claims regarding harming (or benefiting) the unborn. He decides, *contra* Michael Bayles' well-known paper not reproduced here ["Harm to the Unconceived", *Phil. Public Affairs,* 5, 292–304 (1976)], that if it makes sense to say that one can "fail to benefit" an unborn infant by depriving it of birth, it makes equal sense to say that one "harms it," insofar as depriving it of existence constitutes failing to act in its best interests. On these grounds he argues that Bayles' suggestions that sterilization and eugenic control should be legislated in cases where it will prevent birth of infants to whom existence is worse than nonexistence are incorrect; for genetic theory indicates that for a child either to inherit a genetically dominant defective trait, or to inherit a genetically recessive defective trait from at least one homozygote parent, at least one of his parents must share that phenotype, and this constitutes living proof that existence with it is better than nonexistence. Moreover, two heterozygote parents carrying a defective genotype, although not manifesting the phenotype in question, would be prevented by eugenic sterilization from procreating three healthy children for each child inheriting the defective phenotype. To the extent that talk of the rights of possible persons is intelligible, Baker argues, we must conclude that eugenic control violates the rights of the three of the four to healthy birth, even if one could justify depriving the remaining one of existence. Baker also notes the vagueness surrounding the use of the term, "genetic disease."

This vagueness is the major focus of Richard Hull's attention in his paper, "On Getting 'Genetic' out of 'Genetic Disease'." Although all diseases have a genetic component, in that susceptibility, resistance, and injurability are inherited tendencies, those diseases nowadays thought of as "genetic" also seem to be produced by environmental factors, such as diet; and others, for example Tay-Sachs disease, result not from one genetic component alone, but from the combination of this with the lack of another, offsetting character. Thus the delineation of some selected diseases as "genetic" seems arbitrary. Furthermore, theoretic justification is lacking for singling out one of the causal conditions for its incidence as definitive of a disease; the chemical combination of the inner environment seems as good a place to locate "causes" as the genetic makeup. Hull draws our attention to the dangerous consequences of using the term "genetic disease." In suggesting that the genetic condition is determinative, it tempts us to prejudge the issues regarding prevention, control, and cure. This will tend to stifle certain lines of research, and lead to the adoption of false dichotomies; for example, sterilization is frequently thought of as *the* alternative to these diseases, and this assumption leads to the shifting of responsibilities from societal medical institutions to individual persons, and to the stigmatization of those persons who carry the genetic type in question. He reminds us that even the categorization of a condition as a disease presupposes an evaluative comparison with the "norms" supposedly most conducive to a species' "welfare." Care with the vocabulary surrounding health care may help us to minimize prejudice and arbitrary presupposition.

The main essays and the comments on them—which have not been touched on in this Preface—deserve to be read together. They constitute a moral dialog on important issues, a dialog that both layman and professional must enter.

Many current anthologies in biomedical ethics contain, for the most part, articles that have been previously published. The essays in this volume, with one exception, are appearing for the first time. The volume thus contains much new material and gives the reader a good sense of the state of the art. We hope that it will prove as well to be an important contribution to the art.

November, 1978   John Davis
                                                  Barry Hoffmaster
                                                  Sarah Shorten

CONTRIBUTORS

ROBERT AUDI • *Department of Philosophy, The University of Nebraska, Lincoln, Nebraska*

ROBERT BAKER • *Department of Philosophy, Union College, Schenectady, New York*

MICHAEL BAYLES • *Department of Philosophy, University of Kentucky, Lexington, Kentucky*

DAN W. BROCK • *Department of Philosophy, Brown University, Providence, Rhode Island*

MICHAEL D. BUCKNER • *Philosophy and Medicine Program, New York University Medical Center, 550 First Avenue, New York, New York*

COLLEEN D. CLEMENTS • *School of Medicine and Dentistry, University of Rochester, Rochester, New York*

BERNARD DICKENS • *Faculty of Law, University of Toronto, Toronto, Ontario, Canada*

ARTHUR J. DYCK • *Kennedy Interfaculty Program, Harvard University, Cambridge, Massachusetts*

JOSEPH ELLIN • *Department of Philosophy, Western Michigan University, Kalamazoo, Michigan*

M. JOSEPHINE FLAHERTY • *Department of National Health and Welfare, Ottawa, Ontario, Canada*

GLENN C. GRABER • *Department of Philosophy, The University of Tennessee, Knoxville, Tennessee*

BARRY HOFFMASTER • *Department of Philosophy, The University of Western Ontario, London, Ontario, Canada*

RICHARD T. HULL • *Department of Philosophy, State University of New York, Buffalo, New York*

JOHN LADD • *Department of Philosophy, Brown University, Providence, Rhode Island*

ABBYANN LYNCH • *Department of Philosophy, St. Michael's College, University of Toronto, Toronto, Ontario, Canada*

BRUCE L. MILLER • *Department of Philosophy, Michigan State University, East Lansing, Michigan*

LISA H. NEWTON • *Department of Philosophy, Fairfield University, North Benson Road, Fairfield, Connecticut*

DAVID J. ROY • *Center for Bioethics, Institute of Clinical Research of Montreal, Montreal, Quebec, Canada*

SUSAN SHERWIN • *Department of Philosophy, Dalhousie University, Halifax, Nova Scotia, Canada*

DONALD E. ZARFAS • *Department of Psychiatry, Children's Psychiatric Research Institute, London, Ontario, Canada*

CONTENTS

II. ISSUES IN GENETICS

III. THE ROLE OF THE PHYSICIAN

IV. INFORMED CONSENT AND PATERNALISM

V. PROFESSIONAL RESPONSIBILITY

ACKNOWLEDGEMENTS

These papers and the comments on them were presented at a Colloquium on Biomedical Ethics held at The University of Western Ontario in October, 1977. The Colloquium was sponsored jointly by the Faculty of Medicine and the Department of Philosophy, in conjunction with the Faculty of Law. We wish to thank all those at The University of Western Ontario who made the Colloquium possible. Particular mention should be made of Dr. Douglas Bocking, Vice-President, Health Sciences and Acting Dean, Faculty of Medicine; Professor Glenn Pearce, Chairman, Department of Philosophy; Dean David Johnston, Faculty of Law; Dean J. G. Rowe, Faculty of Arts; and the other members of the Colloquium committee: Dr. L. L. de Veber, Department of Paediatrics, Faculty of Medicine and Department of Paediatrics, War Memorial Children's Hospital; Dr. Ian McWhinney, Chairman, Department of Family Medicine, Faculty of Medicine; Dr. Harold Merskey, Department of Psychiatry, Faculty of Medicine; Professor John Snyder, Department of Philosophy and Religious Studies, King's College; and Professor Gregory Brandt, Faculty of Law. The Colloquium was supported in part by a grant from The Canada Council.

In addition, special thanks are owed to Mrs. Maxine Abrams and Mrs. Pat Orphan. Without their cheerful cooperation, diligence, and organizational abilities, neither the Colloquium nor this volume would have been possible.

We also wish to express our appreciation to Mrs. Pauline Campbell, Mrs. Eleanor Sever, Mrs. Penny Switzer, and Mrs. Marion Dundas for help in preparing the manuscript of this volume.

RIGHTS AND MORAL
DECISIONS

Legalism and Medical Ethics

John Ladd

Brown University, Providence, Rhode Island

This essay is concerned with some general questions about methodology in medical ethics. As such it belongs under what is generally called "metaethics" or the "logic of ethics," that is, the second-level inquiry into moral concepts, rules, and principles and their logical interrelations. I believe that it is important to develop an understanding of some of these methodological questions as a propaedeutic to an inquiry into more substantive moral issues. My position, which I have defended elsewhere, is that metaethics cannot, in the final analysis, be separated from normative ethics: for metaethics itself is a normative enterprise and any metaethical theory necessarily has implications for normative ethics, albeit of a very general nature.[1] On the other hand, it is impossible to think clearly and coherently about issues in normative ethics, medical ethics, for example, unless one is cognizant of the logical relationships and implications of the concepts and principles that one is using in analyzing these issues. Much of the purport of this essay will be critical, for I believe that discussions of issues in medical ethics have frequently been on the wrong track; in particular, they have been too legalistic. I shall suggest a new approach, as an alternative to legalism, although what I have to say will necessarily be largely programmatic.

A few general remarks about methodology in medical ethics will provide a setting for what follows. Certain features of the kinds of moral problem with which medical ethics is concerned pose some interesting methodological questions for philosophers. To begin with, moral problems in the medical context are extremely concrete and specific. Their very concreteness and specificity make it difficult to relate them to the

---

[1]See, for example, my "The Interdependence of Ethical Analysis and Ethics," in *Etyka II,* Warsaw, 1973.

abstract theories that have generally been advanced by philosophers, theories such as utilitarianism, Kantianism, intuitionism, natural law and natural rights theories, Rawlsianism, and so on. Some of the difficulties encountered in applying abstract theories such as those just mentioned to concrete and specific subject matters, e.g., particular actions and decisions, were pointed out long ago by John Stuart Mill in his critique of Paley's ethics. Mill noted that the principle of utility is by itself too general and abstract to answer the question: what ought to be done here and now? In order to answer this question, in addition to the principle of utility we need what he called *secondary principles* to mediate between the abstract super-principle and concrete cases of action or decision-making. (Nowadays these principles are called practices, rules, concepts, or moral notions.)

"Without such middle principles," Mill wrote, "an universal principle, either in science or in morals, serves for little but a thesaurus of commonplaces for the discussion of questions, instead of a means of deciding them."[2]

The same sort of consideration applies, of course, to other super-principles such as the Categorical Imperative, the Common Good, Conforming to Nature, or Lessening the Birthpangs of History. As Mill pointed out in his critique of Paley, unless we have regard for secondary principles, we are most likely to go wrong.[3] Although Mill did not stress this point, it should be obvious that secondary principles are context-dependent and so are apt to vary from situation to situation and from society to society. Mill's own theory of liberty illustrates this context-dependence; for that theory can be derived from his utilitarianism only with the help of other premises that are derived from his upper-middle-class Victorian background, premises that he never seriously questioned.

It is no secret, of course, that most moral philosophers, including many utilitarians, have embraced quite uncritically the commonly accepted moral rules and practices of their own society without asking whether and to what extent they are morally adequate. Thus, philoso-

---

[2]From "Dr. Whewell on Moral Philosophy," in J.B. Schneewind, ed., *Mill's Ethical Writings,* New York, Collier, 1965, p. 178. Concerning Paley, he writes in the same passage: "[he] took his leave of scientific analysis, and betook himself to picking up utilitarian reasons by the wayside, in proof of all accredited doctrines, and in defense of most tolerated practices."

[3]Much the same sort of objection as Mill brought against Paley could be brought against those philosophers who attempt to derive answers to specific moral problems from Kant's categorical imperative. To say, for example, that the categorical imperative makes abortion or euthanasia wrong is "picking up Kantian reasons by the wayside."

phers have usually adopted without further question lists of duties, rules, or precepts such as those set forth, e.g., by Ross, Gert, or Donagan.[4] Apart from questions regarding the moral adequacy of these rules and practices, a subject that is not very frequently discussed because they are assumed to be self-evidently true, when we turn to medical ethics we are confronted with quite new kinds of moral problems. For advances in medical technology and revolutionary changes in our social and institutional structures have created many new situations that do not fit so tidily under the secondary principles, rules, and practices of our old morality. Hence, even if our traditional moral categories, rules, practices, and notions are morally acceptable under normal and familiar conditions, they fail to provide intermediate, secondary principles of the sort that are required to address the problems of medical ethics.

In the search for relevant secondary principles and concepts, it has become customary in discussions of medical ethics to turn to the law. Law is obviously designed to deal with concrete cases and thus legal categories, rules, and concepts are by their very nature more easily applied than are such formal and abstract super-principles as the principle of utility or the categorical imperative. Also, because of its flexibility, the law appears better adapted to resolve the new kinds of moral problem that arise in the medical context than are the common sense precepts contained in lists such as those mentioned above; it is obvious, for example, that these moral precepts cannot be applied as easily to coma cases. It is a commonplace of jurisprudence that law, i.e., a law court, is obliged to deal with any case brought before it, either by rendering a judgment or by declaring it outside the court's jurisdiction. For this reason, legal solutions are attractive simply because of their availability, and indeed, precise legal answers are usually available for the moral questions arising in modern medicine. It is the judge's and the lawyer's business to answer such questions. If someone has an answer, even though it is the wrong answer, people are far more likely to seek his or her counsel than that of a philosopher who does not claim to have all the answers, but rather first wishes to examine the questions themselves. Most people, including physicians, would rather have ready made answers than face further questions! In any case, if we wish to dig more deeply we need to ask whether and to what extent it is appropriate to use law and legal categories to resolve the moral prob-

---

[4]See W.D. Ross, *The Right and the Good,* Oxford, Clarendon Press, 1930, pp. 21–22; Bernard Gert, *The Moral Rules,* New York, Harper and Row, 1973; and Alan Donagan, *The Theory of Morality,* Chicago, Chicago University Press, 1977.

lems of medical ethics. The systematic melding of ethics and law is what I refer to as "legalism," which will be explained presently.

The main theme of this essay will be that in dealing with issues of medical ethics, and with the other kinds of ethics that are concerned with problems of modern life, we must begin by recognizing the contextual character of rules, practices, and concepts. Thus, legal answers are relevant in legal contexts, but may not be so in other contexts; by the same token, some moral categories may be appropriate in one context and quite inappropriate in other contexts. In general, then, we must take into account the context in which problems arise and in which our rules, principles, etc., are designed to operate before we can determine how and to what extent they are valid for the radically new kinds of situation that arise in modern medicine and in modern life in general. For this purpose it is also necessary to recognize that moral concepts and rules often operate differently in different contexts and that they perform a multiplicity of functions. The purpose of this essay, then, is to call attention to some of these functions and contexts.

I. GENERAL REMARKS ON LEGALISM

In order to make the discussion of the methodology of medical ethics more concrete, I shall examine in detail one particular approach that I shall call "legalism." By "legalism" I shall mean: "the ethical attitude that holds moral conduct to be a matter of rule following, and moral relationships to consist of duties and rights determined by rules."[5] The side of legalism that interests us here is the use made of the model of law in the formulation, analysis, and solution of ethical issues.

The term "legalism" will be used extremely broadly in this paper to stand for an attitude or a general approach rather than a specific doctrine. I shall argue that the dominance of legalism has distorted our perception of problems in medical ethics as well as our responses to them.

From an ethical point of view, the most significant thing about legalism is that it entails the legalization of morality and the moralization of law, that is, the amalgamation of law and morality. Accordingly, it confuses moral with legal issues, moral reasoning with legal reasoning, and, in general, moral problems with legal problems. Legalistic ethics or jurisprudence of the kind I have in mind

---

[5] I have borrowed this definition from Judith Shklar. See her *Legalism,* Cambridge, Mass., Harvard University Press, 1964, p. 1.

goes back in Anglo-American society at least as far as Blackstone, who thought that the Common Law was, or ought to be, the embodiment of morality.[6]

There are three facets of legalism that should be noted here:

First, legalism takes the task of ethics and of law to be the formulation and establishment of *authoritative rules of conduct,* precepts. Thus, it attributes to law the same function that is generally attributed to ethics, namely, the job of providing answers to questions about what we ought to do. And it requires ethics, on the other hand, to provide rules (precepts, laws) having the same kind of authoritativeness that logically characterizes legal rules, namely, legitimacy, decidability, and enforceability.

Second, legalism favors the kind of argumentation that mixes indiscriminately arguments from authority and consent, the legal model, with arguments from intuitions and general principles, the ethical model. In doing so, it typically ignores some of the distinctive functions of legal argumentation, e.g., its use in litigation.[7]

Third, legalism is reflected in the use of categories of analysis that are partly moral and partly legal. The chief of these moral–legal categories is the concept of rights. Although it would be interesting to dwell on some of the other aspects of legalism, I want to concentrate here on the concept of rights as a typically legalistic category.

Before proceeding, however, I want to emphasize that I do not intend to disparage the law as such. The main point that I wish to make is that the *function of law,* legal argumentation and concepts (e.g., rights), is different from the *function of ethics,* ethical argumentation and concepts. In Wittgensteinian terms, they are two different language games. This assimilation of functions entirely ignores other uses of the law, such as its service as a socially acceptable and effective means of coping with conflicts of interest and of enforcing socially desirable rules and policies. To put it bluntly, we go to lawyers when we are in trouble and not when we need general advice on how to live. Law usually arises out of conflicts, of interests, for example, and provides a social mechanism for resolving conflicts. Ethics, on the other hand, arises out of and is concerned with much wider and deeper perplexities about life and our relations with each other. In general, the failure to bear in mind the essentially different functions of the two kinds of discourse creates a

---

[6]See Gareth Jones, ed., Sir William Blackstone, *The Sovereignty of Law,* Toronto, University of Toronto Press, 1973.

[7]See, for example, Edward H. Levi, *An Introduction to Legal Reasoning,* Chicago, Chicago University Press, 1948.

great deal of confusion, particularly for medical ethics, and makes for
bad ethics and bad jurisprudence.

A few general remarks about legalism are in order at this juncture.
Why is it particularly important to discuss legalism in connection with
medical ethics? There are at least two obvious reasons for doing so.

First, it hardly needs to be pointed out that the most widely
adopted approach to the problems of medical ethics, in this country at
least, is some form or other of legalism. Controversies concerning such
diverse matters as euthanasia, paternalism, experimentation, informed
consent, organ transplantation, and so on are almost always couched
in the language of the law. In fact, it is often difficult to determine
whether the point at issue is a legal or a moral one. For example, is
euthanasia simply a question of what kind of laws there ought to be with
regard to euthanasia or is it, on the other hand, an ethical issue concerning what is right or wrong, independently of what the law is or should
be? Questions about informed consent are ambiguous in the same way.
Indeed, the whole gamut of questions discussed in medical ethics seems
to rest on confusions between law and ethics, in particular, between
legal and moral questions.

Here, however, I want to concentrate on one particular aspect of
legalism, namely, its almost exclusive use of the concept of rights as a
category of analysis for problems in medical ethics. "Is there a right to
—?" is assumed to be the only way to frame ethical questions about
medical matters and "One has a right to—" or "No one has a right to
—" is assumed to be the only way to answer such questions. In fact,
it is hardly an exaggeration to say that discussions of medical ethics
often amount to little more than glosses on the rights to life, liberty, and
the pursuit of happiness. As a result, I shall argue, our view of these
problems and how to go about resolving them is unduly narrow and
dogmatic, *i.e.,* "legalistic" in the worst sense.

A second reason for discussing legalism in connection with medical ethics is a positive one, namely, that legalism provides a particularly
useful method for handling some of the burning issues in the area of
medical decision-making. Many of these issues involve the impersonal
forces of law, professionalism, and bureaucracy, where public and private decision-making intersect and conflict, and where, as I shall show
presently, we usually attempt to solve our problems by appeal to the
law.

Following the contextualist and functional philosophical approach
that I adopt in this essay, I shall begin by indicating some reasons for
the wide appeal of legalism in existing discussions of medical ethics;
here I shall first note some of the worthwhile functions that it serves

in such discussions, and then I shall point out some of its limitations as a framework for medical ethics. Finally, I shall try to show how some of the useful aspects of legalism might be incorporated into a broader and more sophisticated conception of the relationship between ethics and law.

II. THE SPELL OF LEGALISM

The main thrust of legalism, as I have already suggested, is the legalization of morality and the moralization of law. One way in which law and morality are assimilated by legalism is through the notion of *rights*. The use of rights to tie up morality and law is particularly persuasive in our society because, in the Anglo-American tradition at least, the notion of rights has functioned both as a moral and as a legal concept. The concept of rights, however, cannot itself be completely understood without reference to rules of some sort or other, for there is a close logical relationship between rights and rules. The relationship is complex and reciprocal, for, on the one hand, the concept of rights may be used to subsume a claim under a rule or to apply a rule, e.g., such and such ought to be done because X has a right to Y, or, on the other hand, it may be used to justify the creation of a new rule, and so on. I shall argue later that it is quite impossible to understand what is involved in asserting a right without examining the rule, or a possible rule, that it refers to or with which it is correlated in some way. For that reason, often the easiest way to determine what a right (or right-claim) amounts to is to ask for the rule or set of rules that define it and the manner in which it is used. The main point to bear in mind, however, is that legalism constantly shifts back and forth between rights and rules while, at the same time, it assimilates moral to legal rights and moral to legal rules.[8]

The implied relationships between rights and rules and between morality and law may be illustrated by a familiar kind of legalistic argument, namely, that which proceeds from the premise that a certain kind of right exists to the conclusion that there ought to be a law to protect that right. In discussions of euthanasia, for example, we find

---

[8]There is an ancient controversy over whether legal rules are derived from rights or rights from rules. For our purposes, it is immaterial to take a stand on this issue, as long as we recognize the logical relation between the two. I shall return to this subject later. For a discussion of the relationship between concepts and rules, see Julius Kovesi, *Moral Notions.* New York, Humanities Press, 1967.

arguments that start from the premise that there is a right to die and
end with the conclusion that there ought to be a law recognizing that
right as a legal right. Likewise, we find right-to-lifers arguing from the
premise that a fetus has a right to life to the conclusion that a constitutional amendment ought to be adopted that recognizes and protects
that right by outlawing abortion. We find arguments of the same logical
form in ethical discussions concerning the doctor–patient relationship
(e.g., in relation to malpractice, privacy, and informed consent), genetic
counselling and screening, human experimentation (e.g., on children or
prisoners), and so on. Underlying all these arguments is the unsup-
ported assumption that if there is a moral right to X, there ought to be
a legal right to X; that is, legal rules recognizing and protecting moral
rights ought to be created by legislatures and accepted as part of the
law by the judiciary.[9] I shall examine the validity of this assumption
later. In the meantime, given the prevalence of this sort of argument,
it is tempting to ask whether medical ethics is concerned with any
purely ethical issues at all or whether the issues with which it is concerned are not really just issues of jurisprudence, i.e., of what laws we
ought to have.[10]

It is hardly necessary to point out to persons familiar with philosophy that legalism is not a dominant mode of thinking in ethics, either
in the history of ethics or in contemporary ethical theory, although
some recent writings in ethics suggest that it may be becoming more
fashionable. The concept of rights itself was unknown in Greek philosophy, and even in medieval philosophy.[11] Natural rights first came into
vogue in the seventeenth century, where they became the cornerstone
of the political and legal philosophies of such writers as Hobbes, Locke,
and Blackstone.[12] The concept of natural rights was, of course, explicitly repudiated by early utilitarians and, it will be recalled, was labeled
"nonsense upon stilts" by Bentham.[13] Recently, however, there has
been a revival of interest in the concept of rights, partly, as a means of

---

[9]The classical statement of this supposition is found in Blackstone's *Commentaries:*
". . . the first and primary end of human law is to maintain and regulate these absolute
rights, etc." Gareth Jones, *op. cit.,* p. 58.

[10]I shall assume that jurisprudence covers not only theories of what law *is* but also
theories of what law *ought* to be. See Ronald Dworkin, *Taking Rights Seriously,*
Cambridge, Harvard University Press, 1977, chapter 1.

[11]I am told that its first appearance is in the writings of Occam.

[12]See Leo Strauss, *Natural Right and History,* Chicago, University of Chicago
Press, 1953.

[13]*Anarchical Fallacies,* in Jeremy Bentham, *Works,* John Bowring, ed., 1843, vol.
2. Reprinted in A.I. Melden, ed., *Human Rights,* Belmont, California, Wadsworth,
1970, p. 32.

meeting certain traditional objections to classical utilitarianism, e.g., relating to justice, and partly for political and ideological reasons relating to the treatment of dissidents, women, and minorities.[14] Be that as it may, we still must answer the philosophical question: do the concepts of rights and of legalism in general provide the most suitable framework for analyzing the problems of medical ethics?

III. THE UTILITY OF LEGALISM

Let us begin our discussion of the relevance of legalism to medical ethics by noting some of the reasons for the appeal of legalistic, i.e., quasi-legal, approaches to moral and political issues.[15]

At the very outset, it should be observed that legalism has deep historical roots in American society. De Tocqueville remarked on the importance of lawyers in American public life and suggested that the language of the law was the "vulgar tongue" of American politics. He wrote:

> There is hardly a political question in the United States which does not sooner or later turn into a judicial one. Consequently the language of everyday party-political controversy has to be borrowed from legal phraseology and conceptions. As most public men are or have been lawyers, they apply their legal habits and turn of mind to the conduct of affairs. Juries make all classes familiar with this. So legal language is pretty well adopted into common speech; the spirit of the law, born with schools and courts, spreads little by little beyond them; it infiltrates through society right down to the lowest ranks, till finally the whole people have contracted some of the ways and tastes of a magistrate.[16]

What de Tocqueville says about politics applies equally as well to public discussions of almost every general social and ethical problem in American society today, as in the past. The popularity of the legalistic approach in our country might be explained by the fact that in such a heterogeneous society as ours, a society in which persons from diverse cultural, religious, and ethnic backgrounds are forced to live together and to interact with each other, a shared idiom or conceptual frame-

---

[14]See A.I. Melden, *op. cit.* For an up-to-date discussion of the role of rights in our society, see Richard Flathman, *The Practice of Rights,* Cambridge, Cambridge University Press, 1976.

[15]For a more extended discussion of legalism, see Shklar, *op. cit..* and Stuart A. Scheingold, *The Politics of Rights,* New Haven, Yale University Press, 1974.

[16]Alexis de Tocqueville, *Democracy in America,* New York, Vintage, 1959. I, p. 290.

work is required to make public discussions of the important political and social problems that we face together possible. The language of the law serves as a kind of *lingua franca*, as it were, for such discussions by providing a set of commonly agreed upon ground rules for the settling of conflicts and disagreements, as well as commonly accepted premises for use in public debates.[17]

Another practical advantage of law and of the concept of rights in particular is that, insofar as they establish rules of conduct, they serve to define our relationships with strangers as well as with people who we know. For a legal relationship can exist between me and someone who has nothing in common with me except residing in the same legal jurisdiction, e.g., between me and someone who runs into my automobile. A general acceptance of the law has immense practical advantages in a mobile society like ours, for it provides a simple and ostensibly trustworthy means of securing our rights as we travel to new places and meet new people; moreover, it also saves time!

In the medical context, the utility of appealing to rights is even more obvious, for we may find ourselves in a hospital bed in a strange place, with strange company, and confronted by a strange physician and staff. The strangeness of the situation renders the concept of rights, both legal and moral, a very useful tool for defining our relationships to those with whom we must deal. It should be observed in this connection that legalism goes far beyond the narrowly legal, for it provides rules for the guidance of actions in general, rather than merely rules for settling disputes, e.g., in litigation. Thus, in our society, legal categories are used to frame and articulate many sorts of social practices that actually have no legal standing at all.[18]

Yet another useful function of law, and of the concept of rights, is that they define our relationships to such impersonal beings as formal organizations, e.g., hospitals. As I have argued elsewhere, formal organizations are not moral beings and therefore cannot, and should not, be perceived as moral persons having moral obligations, rights, and responsibilities.[19] Obviously, they cannot possess such moral attributes

---

[17]In this regard, law, as a body of accepted rules, could be compared to the rules governing games, for it is a conspicuous feature of games that people from entirely different cultural and ideological backgrounds can play such games as tennis or chess with each other, despite their differences, simply owing to their acceptance of the same ground rules.

[18]Scheingold illustrates this point with the hypothetical example of a faculty debate over a disciplinary issue. See Scheingold, *op. cit.,* pp. 39–49.

[19]See my "Morality and the Ideal of Rationality in Formal Organizations," *Monist* **54** 488–516 (1970).

as compassion, sympathy, or concern. Since they are nonmoral beings, our relationship to them must be defined in nonmoral terms, e.g., in legal or quasi-legal terms; for although they are not moral persons, they do, of course, have a legal existence and are capable of having legal relations, e.g., *legal* rights and obligations. Insofar as American popular ethics is legalistic, legal categories can also be used to define the nonlegal rights and obligations of these impersonal forces.[20] In general, then, legalism is very useful in that it provides a "rational" way for individual persons to exert pressure on nonpersons such as hospitals and insurance companies. For the same reason, it gives us a means for coping with individuals who perceive their relationship to us in impersonal terms, e.g., doctors whose relationship to us is legalistically defined by their professional role.[21]

The importance of this last point should not be underestimated, especially as far as medical ethics is concerned, for one of the most urgent and critical moral questions for modern mass-industrial society is how to reconcile the moral responsibilities of individuals with the increasing power and authority of bureaucracies and other rule-governed groups, e.g., the professions. It is obvious that many of the most perplexing issues of medical ethics arise out of attempts to answer this question. For it is clear that the fortunes, health, and even the lives of individuals are becoming increasingly subject to impersonal decision-making by officials and professionals who represent, e.g., hospitals, drug companies, and the medical profession.[22] This decision-making, in turn, depends for its legitimacy and validity, not to mention its direction, on rules laid down by organizations, e.g., formal and professional organizations, or imposed on them from without by legal authorities or by the marketplace.

Legalism is important in this regard because the authority of impersonal decision-making organs, as well as their limitations, are determined by a network of rules, formal and informal, legal and nonlegal, moral and nonmoral (or even immoral). Of special note are those rules that are adopted by organizations and associations for the regulation of the conduct of their members, officials, and employees, but that are not strictly speaking legal rules, for they are not, as such, created,

---

[20]The denial of the moral personality of formal organizations does not mean, of course, that our relationship to particular persons within an organization need not be moral or that the organization itself might not be externally regulated by moral considerations.

[21]See Richard Wasserstrom, "Lawyers as Professionals: Some Moral Issues," in *Human Rights* (American Bar Association) 5, 1–24 (1975).

[22]See Victor Fuchs, *Who Shall Live?* New York, Basic Books, 1974.

applied, and enforced by organs of the state. For convenience, I shall call them *public rules*. This term will be used as a general label for nonlegal rules of various sorts, including regulations, social norms, conventions, and accepted practices.[23] By calling them rules, I want to emphasize that, like legal rules, public rules are given an explicit formulation; they function as guides to conduct and are used to justify it. Furthermore, like legal rules, they have both formal and informal sanctions for enforcing compliance. If you do not obey the public rule, you may be fired, may not be promoted, may be abandoned, or may not be certified, and so on.[24] Two more points aggravate the situation as far as the individual who is subject to these rules is concerned, e.g., as a patient in a hospital. First, many public rules are unwritten rules; sometimes they are known only to members of the organization itself, e.g., the staff. Second, the rules are imposed on the client–patient–victim from without and without that person's consent. For example, as everyone knows, many if not most hospitals operate under an unwritten rule that nurses are not to divulge a diagnosis or prognosis to patients without the doctor's consent.[25]

It is hardly necessary to point out that many of the most troublesome problems of medical ethics arise in the context of impersonal decision-making authorized and governed by public rules of one sort or another. Sooner or later, almost all of the issues relating to such things as euthanasia, the doctor–patient relationship, confidentiality and record-keeping, the initiation or termination of treatment, the operations of ICU's, and so on, lead to questions about the public rules of organizations such as hospitals, or of the medical profession, e.g., questions concerning which rules ought to be adopted, changed, revoked, overridden, ignored, and so on.[26]

The recognition that we are dealing with a network of rules, or more accurately with many networks of rules, leads to questions about

---

[23]Public rules, in this sense, are quite like what Sartorius calls "social norms." See Rolf Sartorius, *Individual Conduct and Social Norms,* Encino, California, Dickenson, 1975. For reasons that will become clearer later, I do not regard such norms as moral norms, as Sartorius does.

[24]See Charles Perrow, *Complex Organizations,* Glenview, Illinois, Scott, Foresman, 1972.

[25]See letter from George J. Annas to the *Boston Globe,* November

[25]1976, in which he writes: "It is 'standard medical procedures' and archaic hospital practices that promote such misunderstandings of the law by staff members and contribute to patients' ignorance of their conditions." See, also, his *The Rights of Hospital Patients,* New York, Avon, 1975.

[26]See, for example, the rules adopted by Massachusetts General Hospital and Beth Israel Hospital in Boston with regard to the resuscitation of incurable patients in *New England J. Med.* **295,** 7 362–66 (1976).

the logical interrelationships between rules and between systems of rules. For present purposes, it is sufficient to note that one rule may override another in the sense of nullifying it, placing it in abeyance, or giving it less priority. The property of overriding is a logical property and, as such, must be distinguished from the property of being more effective or influential, *i.e.,* of being given greater weight by someone or by some group or organization. Thus, in actual practice, rules about paying for hospital services may have greater weight than rules about providing such services, but we still want to say that the latter are more binding than the former, and that they should override them.

In the hierarchy of rules, legal rules are ordinarily taken to override public rules, such as hospital regulations. (In turn, moral rules are supposed to override legal rules.) Within the hierarchy of legal rules, those derived, say, from the U.S. Constitution have priority over other rules, such as the statutes of state legislatures. Legalism, as an ethicolegal theory, places "constitutional values" or human rights at the top; for, following Blackstone, Americans take for granted that the Constitution embodies moral as well as legal rules. These are the rules of last resort, so to speak.

Underlying this conception of rules appears to be a covert assumption to the effect that a rule can be overridden only by another rule.[27] If this assumption is accepted, and it seems to be a necessary part of legalism, then considerations such as ends, utilities, purposes, motives, and responsibilities are automatically ruled out of court as having no logical power to override public or legal rules, or at least rules of the highest sort. This is an extremely important point, for it shows that some of the ethical problems of medicine have their source in what I take to be a questionable logical (or metaethical) presupposition. I shall return to this point later.

In sum, insofar as legalism is able to provide a method for evaluating, criticizing, and ordering public and legal rules through the use of, say, the concept of rights, then we can agree that legalism provides an answer to many of our needs, social and political, especially in areas where personal considerations are irrelevant or ineffective. In particular, legalism seems to serve a vitally important function in protecting our interests and concerns against the encroachments, not only of government, but also of formal organizations and professional associations. The Lockean concept of natural rights was, of course, originally conceived as a weapon against absolutist government.[28] It therefore

---

[27]We may assume here that rights reflect rules in disguise.

[28]See Sir Ernest Barker, *The Social Contract,* New York, Oxford University Press, 1962, "Introduction."

seems quite appropriate to use it against more modern versions of tyranny.

Nevertheless, before we decide that legalism provides the only acceptable approach to moral problems, e.g., those of medical ethics, we ought to take a second look at the moral bases of legalism and consider some of its limitations. On closer inspection we will find that the posture of impersonality, which constitutes the strength of legalism in one context, is precisely the characteristic that renders it dysfunctional in other contexts. Medical ethics, as I have already pointed out, is specifically—perhaps uniquely—concerned with situations in which impersonal and personal considerations overlap and are difficult to separate from each other. However important the public side is, we neglect the private side only at the risk of abandoning morality and humanity altogether—as legalism sometimes seems to require us to do.

IV. THE LANGUAGE OF RIGHTS

Let us now turn to our main legalistic category, the concept of rights. As already noted, the concept of rights is generally used to provide the link between morality and law (including public rules). Let us begin by asking what is usually meant by the term "right" when it is used in expressions like the "right to life," "the right to die," "the right to refuse treatment," "the right to know one's diagnosis and prognosis," "the right to have an abortion," "the right to medical care," "the right to health," and so on.[29]

We must assume that the rights in question are *moral rights,* sometimes called "natural rights" or "human rights", rather than strictly *legal rights* or probable legal rights, sometimes called "positive rights." Moral are distinguished from legal rights by not being dependent for their existence or validity on creative acts of legislatures, constitutional conventions, or courts, that is, on legislation by human agencies of some sort or other. They are therefore said to be "eternal" or "absolute."[30]

The first question we must ask is: what is the relationship between the two kinds of rights, moral and legal? Legalism characteristically conceives this relationship in very simple terms, namely, that moral

---

[29]For a general discussion of rights, see Flathman, *op. cit.*

[30]"By absolute rights of individuals we mean those which are so in their primary and strictest sense; such as would belong to their persons merely in a state of nature, and which every man is entitled to enjoy, whether out of society or in it." W. Blackstone, in G. Jones, *op. cit.*, pp. 57–58.

rights represent claims that *ought* to be made into legal rights, that *ought* to be protected and enforced by law.[31] In Kant's terminology, a moral (or natural) right is a possible legal right, it is a right *lege ferenda,* so to speak.[32] Thus, civil rights relating to nondiscrimination, before becoming legal rights through an act of Congress, were, as moral rights, rights that ought to be legally recognized. By the same token, advocates of such rights as the right to life or the right to die conceive of them as rights that ought to be recognized by the courts or by legislators.

If we conceive moral rights in terms of this model, then both kinds of rights, moral and legal, have the same content or substance.[33] (Provided, of course, that we are concerned only with those legal rights that embody moral rights.) The chief difference between them is that legal rights are part of a coercive order (Kelsen) and are determined, applied, and executed by public authorities—the courts and the police. Also, unlike purely moral rights, legal rights usually carry specific legal remedies with them.[34]

I shall argue that it is this potential relationship to legal rights that determines the particular role that moral rights play in discussions of medical ethics. It also determines the nature of the arguments that can be given for or against their alleged existence. In passing, it should be noted that the question of the relationship between legal rights and moral rights is more complicated for citizens of the United States, because of the special position of constitutional rights in our political system. As Scheingold points out in *The Politics of Rights,* Americans tend to regard constitutional rights as both legal and moral rights. These "constitutional values" lie at the basis of what he calls the "myth of rights."[35]

For the reasons just given we may expect both kinds of rights,

---

[31]"For the principal aim of society is to protect individuals in the enjoyment of those absolute rights, which were vested in them by the immutable laws of nature . . ." W. Blackstone in G. Jones, *op. cit.,* p. 58.

[32]*Lege ferenda* means literally "as what will or ought to be law." See Kant, *Metaphysical Elements of Justice,* John Ladd, trans., Indianapolis, Bobbs-Merrill, 1965, p. 33. Kant's theory of rights represents a typically eighteenth-century doctrine that the main purpose of the state is to enforce natural rights, e.g., rights to property. On his view, therefore, rights are not, as such, created by the state.

[33]See, for example, Kant, *op. cit.,* pp. 70–71.

[34]There is a time-honored legal maxim to the effect that there is no right without a remedy: ". . . it is a general and indisputable rule, that where there is a legal right, there is also a legal remedy, by suit or action at law, whenever that right is invaded." W. Blackstone in G. Jones, *op. cit.,* p. 139.

[35]See Scheingold, *op. cit.,* chapter 7, esp. p. 94. The common assumption is "that there are satisfying constitutional solutions for all of society's problems."

moral and legal, to have in general the same logical properties. This similarity of logical structure is, I take it, a basic presupposition of legalism. It suggests that, if we want to understand the logical function of the concept of rights in moral discourse, we should look for clues in the way it functions in law. Since the logical properties of rights have been discussed extensively in the literature, and I have discussed them in some detail elsewhere, I shall limit my discussion to three of these logical properties: (1) the peremptory nature of rights, (2) the particular kind of interpersonal relationship implied in the appeal to rights, and (3) the ethical importance of distinguishing between the possession and the exercise of a right.[36]

At the very outset, it should be observed that, as it is generally used, the concept of rights is a very powerful concept. Rights are generally thought to override other sorts of moral considerations. Accordingly, for example, considerations of good will or of charity are subordinate to considerations of rights (or of justice). There is an almost universal, even sacred, commitment to the principle: *rights ought to be respected.* To deny that principle is heretical, indeed, unAmerican! Because of the presumed inviolability of rights, disputes concerning them are transmuted into disputes over whether or not a certain specific right exists, or over the interpretation or application of an agreed on right, or over what is to be done if two rights conflict.

The first property of rights that I have mentioned is their *peremptoriness.* That is, unlike other moral considerations, such as appeals to generosity, appeals to rights are demands that are peremptory; to secure them, it is usually permissible to use coercion, either in the form of legal action or in the form of self-help. In general, when a person's rights are involved, many sorts of action are authorized that would otherwise be impermissible.[37] For this reason, natural rights could be used to justify the American and French Revolutions. On the individual level the right of self-defense, either legal or moral, is used to excuse the killing of

---

[36]For further discussion of the logical and ethical properties of rights, see my "The Definition of Death and the Right to Die" in John Ladd, ed., *Ethical Issues relating to Life and Death,* New York, Oxford University Press, forthcoming, and my "Legal and Moral Obligation," in J. Roland Pennock and John Chapman, eds., *Political and Legal Obligation,* New York, Atherton Press, 1970. See also Joel Feinberg, "The Nature and Value of Rights," *J. Value Inquiry* 4, 243–257 (1970), reprinted in Samuel Gorovitz, *et al.,* eds., *Moral Problems in Medicine,* Englewood Cliffs, Prentice-Hall, 1976, and in A.I. Melden, *op. cit.*

[37]Thus: "... a 'right' is well defined as 'a capacity residing in one man of controlling, with the assent and assistance of the state, the actions of others.' Holland, Jurisprudence 69." *Black's Law Dictionary,* Fourth Edition, St. Paul, Minn., West, 1951, p. 1486.

another person. By the same token, rights are often used to justify medical procedures or nonprocedures that would otherwise be regarded as inhumane.

A second important logical property of rights is that they represent a *relationship* between two persons (or parties): the right-holder and the right-ower.[38] To have a right is to have a right against someone (or against anyone or everyone). The natural and normal situation in which one asserts a right occurs when the person against whom it is asserted threatens, neglects, or otherwise appears unwilling to accede to one's requests, needs, or demands. There is a sense, therefore, in which the assertion of a right is reactive, that is, it represents a response to another person's actual, probable, or possible negative behavior. In sum, the concept of rights is most characteristically invoked in an adversary context or at least in a *possible* adversary context.

Finally, it is particularly important for an understanding of the concept of right to take note of the distinction between *possessing* a right and *exercising* it. Flathman suggests that one can possess a right only if one can choose not to exercise it.[39] In this regard, rights reflect the concept of a person as self-directed or self-governed. Strictly speaking, this condition requires that right-holders (and right-owers) be competent adults capable of self-directed choice. The idea of a proxy who exercises a person's right would therefore be an anomaly, or at least it would have no moral standing; it would still, of course, serve a useful legal purpose.

It should be observed that fortunately people do not choose to exercise most of the rights that they possess: they either intentionally or habitually fail to exercise them, or they may deliberately waive one of them. Whether or not there are any rights that cannot ever be waived is an important and controversial matter.[40] In any case, it is often considered virtuous to refrain from exercising one's rights. Thus, out of charity, a person may waive the right to collect a debt. It should be clear that we will have completely misunderstood the concept of rights if it is assumed that individuals have a moral obligation to exercise their rights. The possibility of choosing

---

[38]When considered as a practice, other parties may be involved, as it were to support or enforce the rights. See Flathman, *op. cit.* This aspect of rights will be commented on in Section 5.

[39]See Flathman, *op. cit.,* pp. 71ff.

[40]Thus, right-to-lifers maintain that the right to life cannot be waived, whereas, on the other hand, those who support the right to die believe that the right to life can be waived and, of course, the right to die can itself be waived.

between exercising and not exercising one's rights, as we shall see, opens the way for such residual ethical categories as love, good will, concern, compassion, and sympathy.[41]

Nevertheless, if any one of these three conditions is omitted or modified when rights are being claimed, the appeal to rights loses its stringency. For example, as soon as the notion of a proxy is introduced, except in the legal context, such rights as the right to refuse treatment lose their persuasive force and efficacy, in part, of course, because of wrangles over who properly may be authorized to decide. If the physician turns out to be the person authorized, the right to refuse treatment becomes an absurdity, because the physician as proxy would have to be able to exercise the right against the same physician who provides the treatment![42] Again, if a right is not regarded as peremptory, then it could hardly be considered a right.

My argument, in what follows, is that these three properties of rights show why it is sometimes quite inappropriate ethically to base medical decisions on the notion of rights alone. For neither medical advice nor patients' requests need be peremptory, that is, advanced as demands backed by force, nor does the doctor–patient relationship need be an adversary one. (This is not to say, of course, that it may not turn into an adversary relationship.)

Quite apart from these considerations that cast doubt on the appropriateness or even the relevance of the appeal to rights in the medical context, many moral problems involved in medical decision-making simply bypass questions about rights, on the general assumption that none of the parties involved feels it necessary to appeal to or exercise these rights. For example, although a patient may have the right to refuse treatment or the doctor may have a right to refuse to treat, it is difficult, if not impossible, for them to discuss rationally with each other what treatment should be undertaken if either of them chooses to exercise such a right.[43]

---

[41]These categories involve, for example, acts of supererogation, which will be discussed in the next section.

[42]It follows from this that, e.g., the right of parents to decide about treatment for their children is a right that they possess as parents and not a right that they acquire as proxies or agents of the child.

[43]I do not wish to suggest that a patient's rights are not sometimes entirely ignored by physicians and hospital staffs when medical decisions are made. Rather, I simply wish to point out that their honesty and honorableness are not requirements based on rights; common decency and respect for the patient are moral desiderata quite apart from any rights the patient may possess. This point should become clearer as we proceed.

The important point for us is that standing on one's rights is a last ditch stand, to be taken only after communication has broken down or when there was no hope for communication to begin with. If a patient cannot communicate about diagnosis or treatment with either the doctor or the hospital staff, or perhaps does not even know who is making the decisions, then it is necessary and appropriate to appeal to one's rights. If one can appeal to a legal right, e.g., to refuse treatment, then that patient is more fortunate than one who can only appeal to a moral right. But even an appeal to a moral right will be more effective than simply appealing to the doctor's good will, sympathy, or understanding.[44] Indeed, according to an ethics of rights, any acts that are over and beyond those required by a right to which a person is entitled are "gratuitous," and are known as "acts of supererogation."[45]

The truth of the matter is that the language of rights is not always the best ethical language for handling morally perplexing situations arising in the medical context. For it hardly seems necessary to point out that there are occasions when the appeal to rights is not only inappropriate, but even immoral. I may have a *right* to refuse you something that you need, e.g., blood or a kidney, but for one reason or another I ought not to do so; to refuse may itself be immoral (irresponsible). Sometimes considerations based on compassion, humanity, or a personal relationship may provide more appropriate reasons for a decision than a reference to rights.[46] In general, requests for help from family and friends are not based on who has a right against whom.[47] Indeed, if one insists on a right, it will more than likely destroy the relationship altogether because it implies the absence of trust, which may sometimes be more important ethically. That is why the appeal to rights is sometimes inappropriate, improper, and immoral.[48]

[44]See Richard Wasserstrom, "Rights, Human Rights, and Racial Discrimination," in Melden, *op. cit.,* pp. 96–110.

[45]See Joel Feinberg, "The Nature and Value of Rights," *op. cit.*

[46]See Bertram and Elsie Bandman, "Rights, Justice and Euthanasia," in Marvin Kohl, ed., *Beneficent Euthanasia,* Buffalo, Prometheus, 1975, pp. 81–99. This article provides some strange examples of the appeal to rights.

[47]I assume that doctors and nurses can be, but hospitals, as nonpersons, cannot be, friends in this sense.

[48]". . . and when men are friends they have no need of justice." Aristotle, *Nichomachean Ethics,* 1155a25.

V. MORALITY AND THE ETHICS OF RIGHTS

It may help us to understand the underlying issues better if we take a closer look at the kind of legalistic conception of morality that takes the notion of rights as its basic moral category. I shall call this conception of morality an *ethics of rights.*[49]

In order to see what is involved in an ethics of rights, we must first ask how the concept of rights operates ethically, that is, how are rights connected with action or human conduct in general? What kind of action does the existence of a right require?

To begin with, a right imposes an obligation on the right-ower to do or to refrain from doing something; accordingly, the concept of a person's rights is commonly analyzed into obligations (or duties) of persons against whom the right avails (i.e., the right-owers). So far, then, an ethics of rights boils down to an ethics of strict obligation.[50] To the extent that it limits obligations to those corresponding to rights, an ethics of rights is, in effect, a minimal morality (a "laissez-faire morality").

Another distinctive feature of the concept of rights is that rights usually, or perhaps always, impose moral requirements on *third parties;* for example, in the case of legal rights they imply obligations on the part of the courts and the executive to enforce them, and in the case of moral rights they require third parties to defend, support, and aid rights-holders in the pursuance of their rights, or at least to avoid interfering when such rights are involved. The third party aspect of rights has a great deal of practical importance, especially in medical ethics, where

---

[49]It might be appropriately called "laissez-faire morality," for it surely has affinities with laissez-faire politics and economics. Many of the points made here are set forth in Joel Feinberg's "The Nature and Value of Rights," *op. cit.*

[50]It is assumed here that rights and obligations are correlatives. Strictly speaking, if we follow Hohfeld and include legal rights in our analysis, we must allow for rights of certain kinds that do not imply correlative obligations, e.g., immunities and powers. But we are concerned with an ethics of conduct, rather than with legal relationships, so that it is unnecessary to include such rights here. We also need to exclude obligations (or duties) to oneself from an ethics of rights, since the notion of a right against oneself is an absurdity. This exclusion has some interesting consequences for medical ethics; for it appears to follow that, according to an ethics of rights, unless it can be shown that the doctor has a corresponding right of some kind, patients do not have any moral obligation to follow their doctors' orders. For further discussion of the concept of rights, see my "The Definition of Death and the Right to Die," *op. cit.* Hohfeld's classical analysis of the concept of right is to be found in Wesley Newcomb Hohfeld, *Fundamental Legal Conceptions,* New Haven, Yale University Press, 1919.

many issues are concerned with the intrusion of third parties into relationships that are essentially bilateral, e.g., doctor–patient, nurse–patient, or patient–family.[51]

An ethics of rights that limits itself to rights and obligations is obviously defective, for, on almost anybody's view, a considerable part of morality is left over after the rights-obligation component has been subtracted, for example, acts of good will, charity, and so on. (From the point of view of an ethics of rights, such acts might be connected with the exercise of a person's rights.) In order to bring acts of this kind into one's conception of morality, it is necessary for an ethics of rights, in addition to recognizing rights–obligations, to make some provision for "moral extras." Traditionally, these "extras" have been called *acts of supererogation.* Acts of supererogation are acts that are "over and above duty"—they have been described as the kind of acts performed by saints and heroes.[52]

For reasons like these, adherents of the kind of legalism that I have called an "ethics of rights" usually divide morality into two parts, one part that is mandatory (i.e., obligatory) and another part that is elective (i.e., supererogatory).[53] The division of morality into these two separate and disparate categories does not seem entirely unreasonable as far as relations between strangers (or enemies) is concerned. For example, if someone meets a stranger in the desert, that individual has the right to refuse to give the stranger a drink of water, especially if it is needed personally. If, nonetheless, that person does give a drink to the stranger, what has been done is something over and beyond the call of duty—it is an act of kindness or generosity. To sacrifice oneself for strangers

---

[51]The involvement of third parties in issues concerning rights is an ethical conception that I am willing to defend against the common legal notion that contracts do not bind third parties. Because of the logical requirement of third party involvement in questions of rights, the reduction of moral relationships to rights-relationships has effectively transferred ultimate decision-making authority to third parties, e.g., the courts, hospital administrations, or insurance companies. Everyone concerned is aware of this development as a practical political or economic problem, but perhaps may not be quite so aware that it is a logical consequence of an ethics of rights.

[52]See J.O. Urmson, "Saints and Heroes," in A.I. Melden, ed., *Essays in Moral Philosophy,* Seattle, University of Washington Press, 1958, pp. 198–216. Also reprinted in Joel Feinberg, ed., *Moral Concepts,* London, Oxford University Press, 1969, pp. 60–73. See also Feinberg's discussion of the subject and the accompanying bibliography in the same book.

[53]See Grice's similar distinction between obligations and ultraobligations. Russell Grice, *The Grounds of Moral Judgment,* Cambridge, Cambridge University Press, 1967.

is an act of supererogation; for, from the point of view of one's rights, it is entirely gratuitous.

VI. ANOTHER MODEL: MORAL RELATIONSHIPS AND DUTIES

The shortcomings of the rights model as a conceptual apparatus for handling moral problems in the medical context make it evident that we ought to look for other sorts of models. For there may be other kinds of relationships besides the rights-relationships, and other sorts of moral categories besides entitlements, that can give us a better understanding of the moral problems that concern us. The basic notion that we need is one that covers such things as caring, as providing for another person's needs—in more general terms, we need an *ethics of giving and receiving*. The ethics in question should provide a moral principle that bases the rightness of giving and receiving, of helping and being helped, on other kinds of relationships than those representing rights or entitlements.[54]

Morality takes on a different character when we focus on relationships that are not simply of the casual or transient sort that exists among strangers.[55] When it comes to personal relationships of a more permanent nature, such as those between friends or between members of the same family or community, the exhaustive and exclusive division of morality into obligatory acts and acts of supererogation that typifies an ethics of rights does not seem appropriate.[56] The morality of taking care of another person with whom one has some kind of personal relationship cannot be reduced either to an obligation corresponding to a right or to a gratuitous favor (e.g., that of a saint or hero).[57] If we include among such relationships the doctor–patient relationship, it should be clear that we are not obliged to say, e.g., that a doctor's taking

---

[54]I mean, of course, relationships of other than a casual or transient sort, e.g., the kind of relationship that Aristotle referred to as friendship, which includes family and colleagues.

[55]Among casual relationships of this kind I should include commercial transactions, e.g., purchases and sales, contracts, and so on, between persons who are otherwise strangers.

[56]See my "The Idea of Community," *New England Journal,* American Institute of Planners, New England, chapter 1, 1 (1972).

[57]See Alasdair MacIntyre, "What Morality is Not," in G. Wallace and A.D.M. Wallace, eds., *The Definition of Morality,* London, Methuen, 1970, pp. 26–39.

special pains to do something for a patient must fall into one or other of these two categories; for in these contexts, optimum as contrasted with minimum concern is neither something that the patient is in a position to demand peremptorily as a right, nor simply an extra kindness on the part of the doctor.[58]

In order to avoid confusing the moral requirements arising out of personal relationships with what I have been calling "rights" and "obligations," I shall refer to them as *duties* to others.[59] One has duties to one's family and to one's children, to one's neighbors, to one's friends and colleagues, and to one's fellow citizens, and, by the same token, doctors or nurses have duties to their patients. Duties, in this sense, are not peremptory, i.e., something that can be demanded as a right. Furthermore, they are the kind of acts that are and ought to be performed out of love, devotion, loyalty, patriotism, and so on. Thus, acts of this kind are always connected logically with virtuous attitudes inhering in the person who has the duty; in this regard, they are like acts of supererogation.[60] On the other hand, unlike acts of supererogation, the performance of these duties is not a matter of choice; they are not "extras," favors; rather they are required as an essential part of the relationship itself; furthermore, as such, they have their source in the

---

[58]Some of the confusion on these matters comes from identifying legitimate expectations with rights; they are, of course, quite different from the logical point of view. Incidentally, it should be pointed out that the use of a contractual model to explain the doctor–patient relationship entirely obfuscates the issue; for the contractual model transforms the relationship into a rights–obligation relationship which, however mutual, omits other morally essential aspects of the relationship. Some of the limitations of the contract–obligation model in general are described in my "Legal and Moral Obligation," *op. cit.* It should be noted that there are other kinds of nonfamilial personal relationships that are comparable, in this respect, to the doctor–patient relationship, for example, the teacher–student relationship. See Paul Goodman, *Compulsory Miseducation and The Community of Scholars,* New York, Random House, 1962, pp. 169–89.

[59]It is important to note that we are concerned here with *moral* duties and not with duties prescribed as part of a job or office or as defined by a social role. Moral duties in this sense arise out of the relationship itself and are not imposed from the outside. That is not to say, of course, that the relationships giving rise to these duties are not culturally determined. On this question, see my "The Issue of Relativism," reprinted in John Ladd, ed., *Ethical Relativism,* Belmont, California, Wadsworth, 1973, pp. 121–24.

[60]But here, perhaps, we are dealing with *philia* rather than *agape.* See Gene Outka, *Agape,* New Haven, Yale University Press, 1972. Outka emphasizes that where agape is concerned "no distinctions of any kind are permitted." (p. 14). He calls this the principle of "equal regard." I am suggesting, on the contrary, that where personal relationships are involved special and unequal regard is required.

recipient's need rather than in the kindness of the giver.[61] The moral difference between these two kinds of helping is illustrated by two different possible answers to the question: "Why did Tom help Ann?" On the one hand, the answer might be: "because Ann needed help," or, on the other hand, it might be: "because Tom is a kind and generous person."

What is required is a moral principle that bases the rightness of giving and receiving, of helping and being helped, on interpersonal relationships of one sort or another. The principle may be formulated schematically as follows: A ought to do X for B because A is related to B *(ArB)* and B needs X *(BnX)*. For example, a mother ought to feed her baby because it is her baby and the baby needs to be fed. The principle of giving and receiving that I have in mind is nicely expressed in the old socialist slogan: *from each according to his abilities, to each according to his needs.*[62] The mother has the ability, the baby has the need. It is hardly necessary to point out the obvious relevance of this principle about abilities and needs to the doctor–patient relationship: the doctor has the ability (*i.e.,* the know-how) and the patient has the need.

For purposes of the present analysis, we can think of the principle of giving and receiving just mentioned as having a more limited scope than that intended by its originators, who wanted to apply it to society as a whole; for in the present context it is meant to hold only between particular persons who are already related to each other in a specific way. In this regard, the principle in question differs from principles concerned with rights, which are supposed to be 'universalizable.'[63] Thus, for example, barring special circumstances such as a contractual agreement, a mother is morally required to feed her own baby, but not other people's babies; if she feeds other babies, she will be performing an act of supererogation. By the same token, a doctor or a nurse is required to provide care for a patient with whom a relationship is already established, but there is always an implied limitation on who *must* be cared for. Thus if a doctor or nurse volunteers to help a

---

[61]Thus, the relationship gives rise to certain "legitimate" expectations on the part of the person in need because it is founded on, or ought to be founded on, mutual trust. No one is more trusting than a baby or in greater need.

[62]I have been unable to ascertain the precise source of this slogan, which was already current in French socialist literature before 1848. It is generally attributed to Louis Blanc, *Organisation du Travail* (1840).

[63]For arguments for the ethical importance of this "un-universalizable" kind of moral principle relating to personal relationships, see my "The Idea of Community," *op. cit.*

stranger, it is an act of supererogation.[64] The relationship of giving and receiving set forth here is therefore *personal* in the sense that it binds particular persons to other particular persons, rather than persons to persons generally.[65]

VII. RESPONSIBILITIES

The various kinds of moral duty that stem from interpersonal relationships can be brought together under the more general heading of responsibility. By "responsibility" I mean a concern that one person ought to have for another person's welfare by virtue of a special relationship that obtains between them. Under welfare should be included such things as a person's security, health, education, and moral integrity.[66] The formula given in the preceding section may be revised to cover the responsibility relationship: A is responsible for B's welfare (i.e., B's health, education, etc.) because ArB.[67] In this sense of responsibility, parents are responsible for the welfare of their children; friends are responsible for each other's welfare; doctors and nurses

---

[64]It should be observed that, for the present purposes, I am disregarding what might be required by the law or public rules of other sorts, e.g., coming to the aid of victims in an accident.

[65]In philosophical jargon, it is not "universalizable." Personal relationships in this sense are not to be confused with social roles, which are defined by the rules determining rights and duties that are attached to roles. Such role-defined relationships might be called "social relations." The view I am presenting here is offered as an alternative to a morality of roles, which, in my opinion, is legalistic. See Dorothy Emmet, *Rules, Roles and Relations,* New York, St. Martin's Press, 1966, pp. 156–66.

[66]When philosophers use the term "responsibility" they usually mean something entirely different from what is intended here. Most of these other uses of "responsibility" are legalistic in intent and effect; in many cases, to be "responsible" simply means to be "liable," "punishable," or "blameworthy." Such uses of "responsibility," in contradistinction to the one in question here, are *post factum,* after the fact. I am concerned here with "before the fact" requirements, not "after the fact" judgments. Furthermore, as I argue elsewhere, responsibility is properly applied only to states of affairs (e.g., consequences of actions) and not to actions themselves. See my "Ethical Dimensions of the Concept of Action," *J. Phil.* **62,** 634–645 (1965), and "The Ethics of Participation," in J. Roland Pennock and John Chapman, eds., *Participation in Politics,* New York, Atherton-Lieber, 1975, pp. 98–125.

[67]For further details about the content of responsibility requirements, see my "The Ethics of Participation," *op. cit.* Unlike the first formulation, which was concerned principally with actions, the new formulation is concerned with states of affairs or "end-states" to be pursued or avoided, rather than with actions per se. I have argued elsewhere that one cannot be responsible for actions per se. See my "Ethical Dimensions of the Concept of Action," *op. cit.*

are responsible for the welfare of their patients, and so on. Furthermore, being responsible is a kind of virtue, and being irresponsible
is a kind of vice; for it is impossible to be a good parent, a good
friend, a good doctor, or a good nurse without taking one's responsibilities for the other seriously, that is, acting responsibly and being
responsible.

Responsibility, as intended here, is a complex moral notion. As
with what I have called "duties," responsibility is not reducible either
to rights or to simple goodwill or benevolence.[68] In order to fulfill one's
responsibility for another person it is necessary to do what is best for
that person in the long run; thus one must reflect carefully about what
is best for the individual concerned. It is always possible, of course, to
make a mistake in judgment, a miscalculation, in deciding what one's
responsibility is or how to carry it out, and so one might actually do
the wrong thing as far as one's responsibility is concerned. In some
cases the mistake may be excusable, e.g., if it was unavoidable, and in
other cases it may be inexcusable, e.g., if it resulted from carelessness.
The important point is that it is possible for a person to act responsibly
and yet, through no personal fault, fail to carry out one's responsibilities.

It is obvious, therefore, that fulfilling one's responsibility for a
person's welfare differs in important respects from fulfilling that person's rights. To do what a right-holder demands may be inconsistent
with one's responsibility to that person; for giving a person what that
person has a right to may in fact not be good for that person at all,
indeed it may be quite harmful.[69] By the same token, when one exercises
what is clearly one's right, one may cause harm to others for whom one
is responsible or, for that matter, to oneself. Much the same sort of
consideration applies, of course, to acts of charity; for a charitable act
towards another may not really benefit that person at all. For these
reasons, therefore, it should be clear that responsibilities belong to a
different moral category from the categories of rights-obligations and
acts of supererogation.[70]

It would be a great mistake, however, to conclude that these other

---

[68]To act responsibly in this sense is neither simply to comply with a right, nor is
it just to perform an altruistic act.

[69]The moral dilemma that this inconsistency between needs and rights creates is
not unfamiliar to doctors.

[70]Accordingly, I reject Foot's bifurcation of morality (or virtues) into justice and
charity. In my opinion, she misses the point of much of medical ethics by ignoring the
morality of relationships and responsibilities. See Philippa Foot, "Euthanasia," *Phil.
Public Affairs* **6**, 85–112 (1977).

categories ought to be be jettisoned altogether, simply because what they require sometimes conflicts with what is required by responsibility. Dilemmas arising from incompatible moral requirements of one sort or another are not peculiar to situations in which more than one moral category is involved; they pervade the moral life and, in fact, sometimes arise within a single moral category. For example, dilemmas may arise from conflicts between promises, between rights, between relationships (e.g., conflicts of loyalties), and between acts of charity. The problem of conflicting moral requirements (i.e., moral dilemmas) is not at issue in the present discussion.

VIII. A COMPARISON OF THE TWO MODELS

My chief purpose here is to point out that there is another moral category that is especially important for medical practice and that is completely ignored by legalism, namely, the category of moral relationships and responsibilities. The importance of this category for medical ethics will become clearer if we examine how it differs from legalistic categories, e.g., that of rights, as a guide to medical decision-making. There are at least four important ways in which this kind of ethics is unlike legalism.[71]

First, the kinds of consideration that are relevant to moral decisions based on moral responsibility are quite different from those that are relevant to decisions based on moral rights. In deciding questions of responsibilities, a much wider range of factors must be taken into account than in deciding questions of rights; a responsible decision may require consideration of such different things as risks and benefits, other relationships, concerns, needs and abilities of persons affected by and affecting the decision. In addition, in order to make responsible decisions it is usually necessary to "weigh" a number of factors against each other; the final decision often requires what we generally call "judgment." Moral philosophers customarily say that such decisions are the outcome of deliberation, reflection, consultation, and discussion.[72]

---

[71]An ethics of responsibility cannot be applied directly to impersonal institutions simply because responsibility, as conceived here, involves a relationship between persons, and formal organizations (e.g., hospitals) are not persons in the sense required. I have set forth in detail the arguments for the position that formal organizations should not to be treated as moral persons in my "Morality and the Ideal of Rationality in Formal Organizations," op. cit.

[72]See Aristotle on deliberation in Nicomachean Ethics, iii, 3; see, also, John Dewey, Human Nature and Conduct, New York, Henry Holt, 1922, pp. 189–209.

Decisions based on rights, on the other hand, are quite different. They do not permit taking into account most of the considerations mentioned, and they do not involve the same kind of weighing, deliberation, judgment, etc., that is called for in cases of responsibility. Indeed, one of the special and distinctive logical properties of a right is that normally, in determining whether to comply with it, one is not permitted to consider any factors other than those directly relating to the status of the right itself and one's ability to do what it requires.[73]

Second, unlike situations in which the issue is simply one of rights, attitudes are an essential ingredient of action in an ethics of responsibility; for it is impossible to conceive of moral responsibility apart from attitudes of concern, of caring, of being solicitous and considerate, and so on. Such attitudes are part of what is required in a responsible action. Indeed, in order even to describe what kind of action is required by a responsibility of a certain sort, it often suffices to refer to one of these attitudes. As far as rights are concerned, on the other hand, a person's attitudes (and motives) are immaterial; the only thing that counts is that one perform the kind of action required by the right.

Third, in a relationship of responsibility there is a certain kind of antecedent inequality between the parties as far as their needs and abilities are concerned; one person needs to be helped, while the other has the ability to help. The rights relationship, on the other hand, is based on the assumption of a certain kind of antecedent equality between the two parties, as is supposed to exist between those entering into a contractual agreement; of course, in actuality, this presumed equality is more often than not fictional.[74] Nevertheless, the notion of "freedom of contract," a basic concept in the ethics of rights, would make no sense if one party were able to impose its will on the other party because of some sort of inequality between them. Thus, where equality in the sense of independence is required by rights relationships,

---

[73]I call this logical property of rights their "opacity." See my "Definition of Death and the Right to Die" *op. cit.*

[74]Consequently, in political philosophy, the contract theory usually makes up a myth of some sort about the existence of an original equality between persons before they enter into the social contract. Hobbes, for example, writes: "Nature has made men so equal in the faculties of the body and mind as that . . . etc." See *Leviathan,* chapter 13. John Rawls also writes: "It seems reasonable to suppose that the parties in the original position are equal . . . [and] the principles of justice [are] those to which rational persons concerned to advance their interests would consent to as equals. . . ." See *A Theory of Justice,* Cambridge, Mass., Harvard University Press, 1971, p. 19. Flathman calls this kind of equality "reciprocity," which means that "participants in the practice readily and regularly switch from one role to another" (*i.e.,* from being a right-holder to being a right-ower). See Flathman, *op. cit.,* pp. 86–87.

it is of the very essence of responsibility relations that one person be dependent on the other in some way or other. In that sense, such persons are not free operators.

The bearing of these conceptual differences between rights and responsibilities on the medical care situation is obvious, for in crucially important ways the patient is likely to be in a state of inequality and dependence *vis-à-vis* the doctor; to speak of rights, e.g., contractual rights, in such cases is an absurdity. Often, a patient is in no position to assert or exercise a right already possessed; newborn infants and persons in coma present the most obvious cases of inequality and dependence. To accommodate cases of this kind to legalism, the notion of a proxy must be invented. As I have already suggested, however, it is doubtful that the notion of proxy has any moral content, although it does and perhaps, for pragmatic reasons, ought to have a legal status. Given the prior acceptance of an ethics of rights, the helplessness of the patient is generally used to justify the assumption by the doctor of the right to make decisions on behalf of a patient. What I am suggesting is that the unreflective acceptance of an ethics of rights in preference, say, to an ethics of responsibility, inevitably leads to moral confusion and irresponsibility of this kind for the simple reason that the ethics of rights rests on the twin assumptions that, from a moral point of view, someone must have the right and that, in the ultimate analysis, rights relationships can only obtain between equals.[75]

On the other hand, the ethics of responsibility implies a quite different sort of equality that is not very obviously a part of an ethics of rights, namely, the equal worth and dignity of individuals, those who are helpless and infirm as well as those who are able and powerful. For an ethics of responsibility requires that all persons involved in the relationship treat each other with equal consideration. Equal consideration here means that help should be matched to needs, rather than to interests, demands, or merit.[76] It also means that the persons involved in the relationship must treat each other with mutual respect and

---

[75] I must reiterate that I am not questioning the necessity or desirability of such notions as that of proxy consent as legal notions. I am questioning only their moral status.

[76] This kind of equality was called *proportional equality* by the French socialists, who invented the slogan already mentioned. To say that the principle of equality requires that everyone be given the same pills or the same books is to make mockery of that principle. It should be noted in this regard that there is a sense in which an ethics of rights, e.g., contractarianism, allows the parties to a contract to treat each other as means to the satisfaction of their respective interests, for it seems to be part of the nature of a contract that it is designed to serve the self-interests of the parties to it. That is not to say, of course, that contractarianism does not presuppose a form of equality of its own. See preceding footnote.

understanding.[77] In other words, persons morally responsible for others should treat them as ends and not as mere means—*all the way through*, as it were, and all the time, rather than just partially and occasionally as is usually the case when morality is reduced, e.g., to contractual relations.[78] One way, perhaps, in which the difference between these two conceptions of equality could be expressed is to say that in an ethics of responsibility equality is a *terminus ad quem*, whereas in an ethics of rights it is a *terminus a quo*. Initially, for example, the doctor and patient may not be equals, but the object of counselling or treatment should be to restore the patient to some sort of equality with the doctor.

Finally, responsibility relationships are dynamic, that is, they change and develop through time as the needs and abilities of the persons develop and their conditions change. One very important way in which they develop is through open discussion, consultation, argumentation, and persuasion. When there is disparity in, say, knowledge or maturity, the responsibility project may become educational, in the best sense. The doctor–patient relationship is itself, in many ways, often an educational relationship involving teaching as much as treating; sometimes, indeed, the teaching may be mutual.[79] A good doctor explains to the patient the nature of the disease, the prognosis, the options as far as treatments are concerned, their risks, benefits, and so on. One of the physician's aims is, or ought to be, to educate the patient and thereby to help the patient to become personally accommodated to the disease and its various implications. That accommodation is a dynamic process and inevitably brings about changes not only in the patient, but perhaps also in the doctor as well.[80]

An ethics of rights, on the other hand, is static and not subject to the kind of changes and development that characterize the ethics of responsibility. For rights are preexistent and predetermined before the decision-process even begins. Thus, the ethics of rights, as such, leaves no room for the kind of mutual education and accommodation that

---

[77]". . . it is of the essence of proper respect that we encourage others to be co-agents, and accept and welcome them as such, each of us, to engage in this enterprise only in ways that are consistent with this attitude." See W.G. MacLagan, "Respect for Persons as a Moral Principle-II," *Phil.* 35,, 294 (1960).

[78]This interpretation of Kant's categorical imperative and of the notion of equality implied by it is set forth in more detail in my "The Idea of Community," *op. cit.* and in my "Egalitarianism and Elitism in Ethics," *L'Egalité* 5 (1977).

[79]This is one of the main points made in Eric Cassell, *The Art of Healing,* New York, Lippincott, 1976.

[80]Of course, the relationship itself also often changes, as when a baby grows up and no longer needs to be cared for and as the parents grow older and, in turn, become dependent on the child.

may be necessary to change the situation for the benefit, say, of the patient.[81]

Obviously much more needs to be said about an ethics of relationships and of responsibility. But a fuller exposition of that kind of ethics cannot be undertaken in this essay. My only purpose here has been to suggest a possible alternative to an ethics of rights, simply in order to bring out some of the limitations of that kind of legalistic ethics.[82]

IX. THE VINDICATION OF RIGHTS

In the final analysis, it is impossible to understand clearly how rights are related to ethics in general until we face up to the most critical question concerning them, namely: how are these rights vindicated?[83] How can the existence of a moral right be established? How can it be shown that persons in general or particular persons possess moral rights of a certain kind? Questions such as these are especially critical for legalism, for, despite all the affirmations of absoluteness and certainty concerning rights, it is hardly necessary to point out that fierce controversies rage over the existence of certain kinds of right, for example, the "right to life" or the "right to have an abortion." How do adherents of such rights propose to vindicate their claims and to refute the claims of others with whom they disagree?

Obviously, the kind of vindication of a right that can be given depends upon whether the right in question is thought to be basic and underived, or whether it is thought to be derived from another more basic right or moral principle, such as, for example, the principle of utility or the categorical imperative. Legalists usually assume that

---

[81]The reader should not conclude that I am arguing here for some form of paternalism; for unless it is assumed that educational efforts are by definition "paternalistic," it does not follow that in giving counsel and advice one is acting paternalistically. Indeed, precisely the opposite is the case, for in taking the trouble to convince someone to do something or other, one is treating that person as a person worthy of respect and in that sense as an equal. Of course, if paternalism is a purely legalistic concept, as seems likely, it has no use in the present context!

[82]It should be noted, however, that the points that I have mentioned in connection with an ethics of relationships and responsibility bring out the limitations of many other standard kinds of ethical theory, e.g., the theory of justice as fairness (Rawls), utilitarianism, and various types of deontological ethics.

[83]I shall use the term "vindicate" to refer to the rational process of establishing a title. I think this is what Kant had in mind by the term "deduction." For reasons that should be obvious, the existence of a right often cannot be proved deductively, although it can be established by other means.

rights are basic, rather than derived from other kinds of moral principles; in other words, they cannot and need not be "proved." Thus, it
is simply taken for granted that there are such rights as the right to life
—no questions asked.[84]

Accordingly, the most typical legalistic approach to the question
about vindicating moral rights is to suppose that their existence is
self-evident. It is simply assumed that anyone who fails to acknowledge
the existence of an alleged right is defective in some way or other. The
failure is attributed to the person's being morally blind, ignorant, uned-
ucated, or delinquent. Such a person is invariably treated condescend-
ingly, if not ostracized altogether, by the right's advocates. As I have
argued elsewhere, this kind of dogmatism, exemplified among philosophers by the appeal to self-evidence (i.e., intuition), is not only self-
defeating, it is also immoral.[85]

A less extreme approach to the vindication of particular rights is
to attempt to derive such rights as the right to life, to death, to privacy,
and so on, from a single overarching right, such as a general right to
self-determination.[86] But there are difficulties with general rights such
as these if they are regarded as moral rights; an uncritical and inflexible
adherence to the right of self-determination without reference to the
special context in which it is valid is, as suggested earlier, both unrealis-
tic and irresponsible. Does a child have the right to refuse to take
medication? Is there a right to die, if one so desires? Is there a right to
refuse treatment, if one so wishes? If these questions are treated as
moral rather than legal questions, the answer is often: "No." At least,
it is not so obviously: "Yes. It is up to you: you have the right to do
any or all of these things because of your (moral) right of self-determination."

---

[84]Sometimes the mere fact that the Declaration of Independence or the U.S.
Constitution asserts that something is a right is taken as sufficient to establish that it
is a (moral) right!

[85]See my "Positive and Negative Euthanasia" in Michael Bayles and Dallas High,
eds., *Medical Treatment of the Dying: Moral Issues,* Boston, G. K. Hall, 1978.

[86]This right is sometimes called the right of "autonomy." See H.L.A. Hart, "Are
There Any Natural Rights?" *Phil. Rev.* **64**, 175–91 (1955). For comments on this
article, see Flathman, *op. cit.,* p. 234, no. 8. Kant sets forth a similar general principle
of right (or law) in his *Metaphysical Elements of Justice, op. cit.* A careful reading of
Kant's other ethical writings, e.g., the *Foundations of the Metaphysics of Morals,* will
show that he conceived of 'autonomy' primarily as the source of duty, *i.e.,* self-
legislation, and only indirectly as the source of a right. For Kant, autonomy means the
capacity to be moral and not the authority to do anything one wishes to do. It is
unfortunate that the word "autonomy" has been used in so many different senses that
the original Kantian concept of autonomy has become lost in the verbiage.

In regard to self-determination itself, it should be noted that slavish adherence to the principle of self-determination, without regard to the context, simply results in transferring the moral issue to another place and translating it into another set of questions, namely, questions about the competence of an individual to make such decisions personally. This kind of legalism, then, turns questions about dying and suicide, and questions, e.g., about informed consent, into entirely different and perhaps irrelevant questions about a person's competence. This move takes us far away from the original moral question: is it morally right to undertake (or to refrain from) certain acts, e.g., a treatment? Because of the obvious objections to the unqualified principle of self-determination, I prefer to put in its stead what I shall call the principle of moral integrity, which will be explained in the next section.

Another approach to the vindication of rights is to attempt to derive them from moral rules of some kind. For example, the right to life could be derived from the rule against killing innocent persons. It is impossible to discuss this approach here; to do so would require a detailed examination of various sorts of rule moralities. The most well-known of these moralities is rule-utilitarianism. Rule-utilitarianism, as well as some other types of rule morality, usually conceives of moral rules as analogous to legal rules (or institutional rules of some sort); the objection to such rules is that they are customarily framed with a view to general rather than individual cases: "What if everybody did that?" is regarded as a test of the validity of a purported rule. Thus, it is often argued that if there were a general rule permitting euthanasia, then many people would be put to death against their wills. Such an argument might possibly have some cogency when the issue is one of making laws, but it is difficult to see what it has to do with the morality of euthanasia in particular cases. Arguments like the so-called "slippery slope" argument against euthanasia abundantly illustrate the slippery character of legalism at its worst, that is, the approach to ethics that gives legal arguments for moral conclusions and that substitutes moral conclusions for legal arguments.

X. INTEGRITY, AUTONOMY, AND RIGHTS

All these, and other attempts to vindicate rights as the basic category for moral decision-making, e.g., medical decision-making, are bound to fail because they abstract the concept of rights from the context in which it is designed to function, namely, the context of politics, law, institutions, and public life in general. To take rights as providing general guidelines for conduct outside of these contexts leads, as I have suggested, to irresponsible behavior, if not, indeed, often to a cruel and callous attitude towards others in need. Simply to give in to a person who refuses something needed, say, to stay alive may, in some cases, amount to an abdication of moral responsibility.

As I have already pointed out, however, we cannot dispense with the concept of rights (i.e., moral rights) altogether; for the fact is that much of our life, especially when we are sick, is controlled by external and impersonal forces of the sorts already mentioned. In such contexts, our only ethical (and legal?) defense is provided by the concept of rights. How, then, is the category of rights related to the category of relationships and responsibilities?

I can only briefly indicate the answer to this last question. The position that I am prepared to defend is that the concept of rights, as a cluster of claims on society and its institutions on the part of the individual, derives its principal moral warrant from the concept of moral integrity. This concept, unlike the concept of simple self-determination, focuses on the integrity of personal relationships, concerns, and responsibilities. Everyone in society has a duty, individually and collectively, to defend, support, and nourish these moral relationships both personally and in others. The concept of rights provides an effective social and conceptual instrument for carrying out this general duty. On my view, then, the value of legalism and of rights is not intrinsic, but instrumental to the moral integrity of individuals in their personal relationships.

The lessons for medical ethics should be clear. For reasons that I have indicated, we cannot dispense with the concept of (moral) rights; it is a necessary part of our moral armory. Accordingly, we should insist, for example, on the right to informed consent, But we must be careful not to put the cart before the horse—to confuse the instruments of morality with morality itself; for rights themselves are derived from and receive their moral substance from individuals in their relationship with one another. It follows that doctors and nurses, if they wish to relate on a moral basis to patients in their care, cannot construe their relationship purely in terms of a legalistic framework of mutual rights and obligations; if they wish to relate to them morally, they must dig more deeply into questions of their human relationships to and responsibilities for their patients as persons, disregarding if need be the professional and institutional pressures that threaten the integrity of these relationships and of the persons involved in them.

An example may show how rights of this kind can be derived from the principle of moral integrity. Most hospitals prohibit children from visiting their parents when they are in the hospital; when the parent is dying, the inhumanity of such a prohibition is especially striking. That this kind of prohibition can be forced upon parents and children constitutes a direct affront to the integrity, the moral integrity, of the relationship between them. It is not too much to say that dying patients have the right to see their children and that the children have the right to visit a dying parent. In the old days, when people died in their homes, this sort of problem would not have arisen—and the assertion of such a right would have been strange and irrelevant. But today, given the circumstances in which people are forced to die, the affirmation of this kind of right seems not only reasonable but also absolutely necessary —for moral reasons. And as a moral right, with a moral basis, it justifies every effort on our part to have it recognized as a right in law and in the public rules adopted by such institutions as hospitals.

2

Comments on "Legalism and Medical Ethics"

Barry Hoffmaster

University of Western Ontario, London, Ontario, Canada

Professor Ladd's paper is an important and much needed investigation into the theoretical underpinnings of medical ethics. I agree wholeheartedly with his criticisms of a general ethics of rights. My remarks, therefore, will focus on three issues: the connection between legalism and an ethics of rights, Professor Ladd's criticisms of a particular version of an ethics of rights, and the adequacy of the alternative ethics of responsibility that is offered in its place.

Professor Ladd's putative target is a view that he calls "legalism." The concept of rights, however, bears the brunt of his criticisms. Since the rights under attack are moral rights, not legal rights, it is not clear how these criticisms pose any threat to legalism. Legalism, for Professor Ladd, seems to involve several different, but related, points.

1. Use of the term "legalism" emphasizes that problems in medical ethics often are discussed in terminology borrowed from the law. The notion of rights, which is quite prominent in medical ethics, has its home in the law. H. L. A. Hart has made the same point about the concepts of duty and obligation.[1]

2. Legalism is a normative view about the relationship between law and morality, more specifically, the view that legal and moral questions *ought* to be assimilated. Since the concept of rights is common to law and morality, it is easy to make this identification. If one then wishes to know whether active euthanasia, say, is *morally* wrong, one asks whether there is, or ought to be, a law that prohibits active euthanasia. A more formal way of putting the view is: A moral right to X exists

---

[1] See H. L. A. Hart, "Legal and Moral Obligation," in A. I. Melden, ed., *Essays in Moral Philosophy*, Seattle, University of Washington Press, 1958, pp. 83–84.

if and only if a legal right to X exists or ought to exist. As Professor
Ladd notes, this ethical position reduces all questions of medical ethics
to questions of what laws ought to exist, and thus encourages certain
kinds of arguments, *viz.*, rule utilitarian and slippery slope arguments.

3. Legalism is the use of a model of rules to frame and solve moral
problems. Because there is a logical connection between rules and
rights, rights play a central role in such an ethical theory. But again the
rights are moral rights, not legal rights.

These three points are innocuous. As long as one is working with
a coherent notion of moral rights, the pedigree of the notion is unimportant. The assimilation of legal and moral questions is a common, but
nevertheless egregious, mistake, and Professor Ladd must be com-
mended for making people aware of this mistake. Finally, the use of
rules to handle moral issues has a long and distinguished history.

Professor Ladd's criticisms are directed against a particular concept of moral rights, *viz.*, moral rights that share the logical properties
of legal rights. The three logical properties that he considers are: 1. the
peremptory nature of rights; 2. the impersonal relationship presupposed by rights; and 3. the distinction between possessing and exercising a right. If legalism is understood as a moral theory that holds that
rights with these three logical properties play a central role in handling
moral issues, then Professor Ladd's criticisms also apply to legalism.
But the question of whether the concept of rights that is so popular in
medical ethics is a legalistic concept of rights remains open. Do most
people who appeal to rights to handle problems in medical ethics in fact
accept the view that the only moral rights that exist are ones that ought
to be protected by law? Perhaps most people have a different justification, religious for example, for their rights-claims.

Professor Ladd has two main objections to an ethics of rights. The
first objection applies to any ethics of rights, not just to a legalistic ethics
of rights. Any ethics of rights must solve certain internal problems. It
is not always clear how a rights-claim is to be interpreted or applied,
or how conflicts between rights-claims are to be resolved. How does one
decide whether a right to life or a right to control one's body takes
precedence? More important, however, is the problem of vindicating
rights-claims. How do we know whether someone really does have a
right to die, or a right to privacy, or a right to be a bigamist? Do
animals, dolphins for example, have rights? An adequate theory of
moral rights must explain how rights-claims are justified.

The second objection is directed at a legalistic theory of rights.
There are, according to Professor Ladd, situations in which an appeal
to a legalistic moral right is inappropriate or improper, or even im-

moral. He notes that a patient's request need not be a peremptory demand and that the physician–patient relationship need not be adversarial. He also observes that rational discussions of treatment would be impossible if a patient were to exercise the right to refuse treatment. I agree with these points, but I fail to see their relevance. The assumption behind them seems to be that every interchange between a physician and patient is moral, and thus if an ethics of rights is not applicable to every one of these exchanges, it is defective. Although I am in favor of physicians becoming more aware of the many evaluative decisions they make under the guise of medical or scientific decisions, I do not see how Professor Ladd, with these examples, has raised any moral issues at all, much less moral issues to which an ethics of rights is inappropriate. In other words, I do not see what Professor Ladd finds "morally perplexing" about the examples he gives. Moreover, his criticism that an ethics of rights sometimes is immoral begs the question because it presupposes that certain competing moral considerations, *viz.,* those of his own ethics of responsibility, are relevant and overriding.

The overall strategy behind this objection also is puzzling. It is perhaps true that rights-claims are most psychologically appealing, most effective, and most often advanced in impersonal, adversarial, and even hostile settings, such as when one must deal with a stranger or an institution, but these are all empirical claims, and nothing follows from them about the moral appropriateness of rights-claims in other kinds of settings. Likewise, Professor Ladd's assertion that when one bases a moral request to a friend or relative on a rights-claim, such a request implies the absence of trust and therefore is likely to destroy the relationship, is an empirical claim, for which Professor Ladd offers no evidence. To point out that rights-claims are most successful or most effective in certain kinds of contexts says nothing about the moral appropriateness of rights-claims in any context. I find it perfectly appropriate for a child to request that his parents not enter his room without knocking first, and to justify this request by appealing to a right to privacy. Yet this is not a relationship between equals; the request is not a demand backed by force; and the relationship is not adversarial. What this example suggests is that one can have a coherent theory of moral rights that is not legalistic.

I turn now to some problems with Professor Ladd's alternative ethics of responsibility. There are two purposes that one might expect a moral theory to fulfill.[2] One is to provide an account of right-making

---

[2]See R. Eugene Bales, "Act-Utilitarianism: Account of Right-Making Characteristics or Decision-Making Procedure?" *Amer. Phil. Quart.* **8,** 257–265 (1971).

characteristics; that is, a moral theory is to provide an account of the characteristics that all and only right actions have in virtue of which they are right. The other is to provide a method for choosing, from a set of alternative actions, the action (or actions) that in fact is morally right. Professor Ladd's ethics of responsibility fulfills neither of these functions.

Two notions appear crucial to an ethics of responsibility: attitudes and personal relationships. Attitudes can be accommodated within traditional moral philosophy, but not in the way that Professor Ladd wants. A distinction commonly is drawn between questions about the moral rightness and wrongness of actions, and questions about the moral praiseworthiness and blameworthiness of agents. Attitudes are generally held to be relevant to an assessment of an agent's praiseworthiness or blameworthiness, but not to a determination of an action's rightness or wrongness. Professor Ladd, I take it, does not accept this distinction. The moral rightness of an action for him is logically connected with certain attitudes. But he does not specify either how attitudes are right-making characteristics of actions or how attitudes lead to determinate conclusions about the moral correctness of actions. He has not shown, for example, how concern for the welfare of a person with whom one has a special relationship, or any other "virtuous" attitude for that matter, leads to conclusions about *what* actions are morally right or wrong or *why* these actions are morally right or wrong. The same point can be made about the notion of a personal relationship, which is the crucial element in Professor Ladd's principle of giving and receiving. No conclusions about what actions are morally right or wrong follow from the mere existence of a personal relationship. Further, the existence of a personal relationship does not appear to be *in itself* a right- or wrong-making characteristic. Consider the following example of moral argumentation:

You ought not lie to your mother.
Why not?
Because she is your mother.

The answer, "Because she is your mother," does not *in itself* provide a moral reason for not lying. Other reasons can be provided for not lying to one's mother, but the point is that these are independent moral arguments. The mother–child relationship points to these other reasons, but the existence of the relationship itself does not provide a moral reason. One may have special duties or obligations because one stands in a certain relationship to another person, but these duties or obligations are justified on independent moral grounds, not on the mere existence of the relationship.

Professor Ladd's ethics of responsibility is tied closely to the notion of a personal relationship, which he wants to distinguish from social or role-defined relationships. So even if Professor Ladd can handle the above problems, he still must show that the physician–patient relationship qualifies as a personal relationship. There are crucial differences between the physician–patient relationship and more intimate relationships such as husband–wife and parent–child that must not be ignored. One must never forget that being a physician is, above everything else, an occupation and a profession.

Two additional worries with an ethics of responsibility deserve mention. First, it is unclear whether an ethics of responsibility is intended to displace or to supplement a morality of obligation and supererogation. At times Professor Ladd says he is offering an alternative ethic, but elsewhere he concedes that the concept of rights is indispensable to morality. If an ethics of responsibility is an alternative ethic, one needs some idea of the situations in which an ethics of responsibility is appropriate and the situations in which an ethics of rights is appropriate. Professor Ladd suggests a possible criterion when he distinguishes personal relationships and social roles, but, as noted above, this distinction needs to be worked out. Social roles seem to pervade even those relationships that Professor Ladd views as personal. If an ethics of responsibility is a supplemental ethic, the proliferation of moral categories that results makes the resolution of moral problems even more difficult in principle. Professor Ladd is aware of the dilemmas created by competing and apparently incompatible moral considerations. His view makes these dilemmas appear even more intractable.

The second worry is that an ethics of responsibility seems to have a paternalistic bias built into it because it emphasizes the concept of needs rather than the concept of interests as does an ethics of rights. Each individual is the best judge of self-interests; these interests, like tastes, are subjective. But needs are objective, and everyone has roughly the same needs. Thus one is in a position to judge the needs of others in a way in which one is not in a position to judge the interests of others. I know that you need food, water, sleep, sex, love, and so on just as I do. An ethics of responsibility, by stressing the concept of needs, invites one to make judgments on behalf of others. I do not want to claim that paternalistic acts are *ipso facto* morally wrong, or even that they are *prima facie* morally wrong. I merely want to observe that an ethics of responsibility encourages paternalism in a way that an ethics of rights does not. Professor Ladd seems to embrace this consequence when he says: "In order to fulfill one's responsibility for another person it is necessary to do what is best for that person in the long run; thus one

must reflect carefully about what is best for the individual concerned"
(p. 26).

In conclusion I want to offer an alternative diagnosis of what is
wrong with medical ethics. It is interesting that Professor Ladd, despite
his criticisms of an ethics of rights, wants to preserve the practical
merits of rights-claims. Rights are valuable, he admits, because they
protect one from the impersonal decision-making of institutions. A
rights-claim is a strong moral counter-punch to an argument that rests
on, say, what the rules of the hospital require. Professor Ladd finds
attractive the moral protection that rights-claims provide. This ambivalent attitude towards rights suggests an alternative course for medical
ethics. The focus of medical ethics should be the institutional arrangements that make an ethics of rights an indispensable weapon in a
patient's moral armamentarium. One needs to rise above the individualistic, case-oriented approach to medical ethics, an approach which
Professor Ladd's ethics of responsibility exemplifies, and address the
more global questions of institutional design. Professor Ladd recognizes
that many of the troublesome issues in medical ethics are generated by
impersonal decision-making. The crucial question then is: how can one
change the impersonal decision-making processes of institutions so that
morally sensitive decisions emerge? Concentrating on reactive, individualistic decision-making will not improve matters significantly.
The outcome of individualistic moral decision-making will remain un-
satisfactory as long as the institutional contexts in which these decisions
are made are impersonal, intimidating, and irresponsive. Progress in
medical ethics requires that questions of institutional design be addressed.

3

The Moral Rights of the Terminally ILL

Robert Audi

University of Nebraska, Lincoln, Nebraska

The care of dying patients is a major concern of medical ethics. There are problems about how to keep up their morale; problems about how to reduce their suffering in a way that does not dull their remaining time; and moral problems about what treatment to undertake or terminate, and what information about it to give the patient and family. My main concern here will be with these moral problems. I shall explore them by examining certain of the important moral rights of people who are dying, but I shall also consider some of the moral rights of physicians and of a patient's family. Of course, the moral issues raised by terminal illness may be approached differently, for instance by exploring the relevant moral principles; and the same substantive results might be reached by a quite different route. But talk of rights is pervasive in current discussions of these issues; and I believe that by clarifying certain of the rights of terminal patients, their doctors, and their families, we should achieve some insights into how a person in any one of these groups should treat people in each of the others. We should also learn something about the physician–patient relationship in general, and perhaps something about rights, which are not nearly so well understood as one would expect from the frequency and confidence with which many people invoke them to support their moral judgments.

I shall first consider the distinction between killing and letting die, with a view to ascertaining whether it is morally significant, particularly for the issue of euthanasia. My second topic will be the rights of terminal patients regarding their medical treatment; my third, their rights to information about their condition; and my fourth, some legal implications of what I have said.

43

I. KILLING AND LETTING DIE

Euthanasia, when thought of without qualification, is usually conceived as "mercy killing." On the other hand, recent writers on euthanasia commonly distinguish two forms of it: active and passive. Active euthanasia is, roughly, killing in order to spare the subject suffering or "indignity"; and passive euthanasia is, roughly, letting the terminal patient die for the same purpose(s). I shall argue that there is a morally significant distinction between killing and letting die; I thus propose that we *not* use the term 'euthanasia' or even 'passive euthanasia' to refer to letting die. Otherwise the connotations of 'killing' that attach to 'euthanasia' when it is understood as 'mercy *killing*' may unwarrantedly affect our attitude toward letting terminal patients die.

What is the distinction between killing and letting die? I shall not try to define either term rigorously, but this much seems clear: killing a person *(S)* entails doing something which brings about that person's death, whereas permitting someone to die does not entail this, but is, roughly, doing something, or omitting to do something, and thereby letting some already existent process bring about the person's death. You cannot let someone die if there is not *already* something killing that person, something that will kill the person if nothing intervenes. Killing does not entail this condition. Secondly, if an agent *X* (alone) kills *S,* some fatal act of that agent, e.g. an injection, is the cause of, or at least causes, *S*'s death; but if *X* lets *S* die, it would be wrong to say that *X*'s act of, e.g., withholding a drug, is the cause of death. Indeed, even if *X*'s letting *S* die *consists* mainly in a "positive act" like unplugging a machine, it would be a mistake to say that this act caused *S*'s death. To be sure, unplugging a machine attached to a *curable* patient might both cause the death of, and kill, the patient. However, in the usual terminal cases, the cause of death would be not any discontinuance of treatment, but the *disease* or other condition that was killing the patient and *of* which the patient died.[1] That one can kill *or*

---

[1] It is not always easy to determine what caused a death. Suppose a diabetic managing very well on insulin dies because *X* withholds it. If we judged by the autopsy report alone, we might conclude that diabetes caused *X*'s death. This is true, but it is better and equally true to say that *X*'s withholding the insulin caused it. This seems to be mainly because the disease was not killing the victim, which implies that withholding the drug is not letting *S* die. It is killing, and that entails causing death, though not being *the* cause of it, as cases of overdetermination show. Now imagine that the insulin had only a 50–50 chance of saving *S.* Would *X*'s withholding it be the cause of *S*'s death if *S* dies afterward and the autopsy report is precisely the same? The answer is not clear, and this seems to be true at least in part for the same reasons that render it unclear whether *X* kills or lets *S* die. If the distinction between killing and letting die is related in the way this suggests to that between being the (or a) cause of death

let die by unplugging a machine, then, shows that the distinction between killing and letting die cuts across both the distinction between positive and negative actions and that between commissions and omissions. This point has apparently been missed by Fletcher,[2] among others.

Supposing the distinction between killing and letting die is real, what moral significance does it have? There are many possible interpretations of the attribution of moral significance to the distinction, and it is often unclear which of these is intended when the distinction is appealed to in support of moral judgments. The following are some plausible interpretations of the claim that the distinction between killing and letting die is morally significant:

(1) There is a duty not to kill people, and a duty not to let them die; and the former is more stringent.

(2) People have a right not to be killed, and a right not to be allowed to die; and the former is stronger than the latter.

(3) Killing a person, unlike letting one die, is prima facie wrong, in the sense that there are always good, though overridable, reasons to think it wrong.

(4) Other things being equal, killing a person is worse than letting one die.

(5) Other things being equal, if letting a certain person die in a particular situation would be wrong, killing one in that situation would be worse.

---

and not being one, then the fact that the latter distinction apparently has moral significance suggests that the former does also. Letting S die, unlike killing S, does not entail being the (or a) cause of S's death. Indeed, letting S die (without performing any further act) entails the negation of this.

[2]Fletcher asks what is the moral difference between doing nothing to keep a patient alive and giving a fatal dose of a painkilling or lethal drug. "The intention is the same either way. A decision *not* to keep a patient alive is as morally deliberate as a decision to *end* a life." See Joseph Fletcher, *Moral Responsibility: Situation Ethics at Work* (Philadelphia: Westminster Press, 1967), p. 150. Fletcher appears to think that letting die is commonly thought to be a negative action, and rightly points out that it may be "morally deliberate." But it does not follow that it is not morally different from killing. Perhaps Fletcher would grant a moral difference if he agreed that the relevant intentions are not the same. In both cases there may be an intention to hasten death, but that is beside the point.

(6) There are certain pairs of act-types such that (a) other things being equal, one of the acts would be morally preferable to the other, and (b) a sufficient condition for this is that the former is letting a person die, and that the latter is killing one.

It may be that some of the above are equivalent, e.g. (1) and (2), or (5) and (6). However, nothing I say will turn on this question. Items (1)–(6) all stand in need of analysis. I shall not attempt to analyze any of them here, but I believe that by considering several examples we can clarify them to some degree and develop evidence regarding which, if any, are true.

It may be useful to begin by imagining two worlds that are like ours, apart from whatever differences are implied by the following conditions: in one of them people obey a principle prohibiting killing, but have no principle prohibiting letting die; the other is identical except that people obey a principle prohibiting letting die, but have no principle prohibiting killing. If it is not more important to prohibit killing than to prohibit letting die, we should find no significant difference. But there is a huge difference: in the world in which killing is allowed, life would doubtless be, in Hobbes' phrase, "solitary, poor, nasty, brutish and short." In the world in which letting die is allowed, there would be some heartless refusals to render aid, but those not afflicted with terminal conditions, or blessed with supportive family or friends—most people, presumably—could manage.

One might reply with two objections: first, that the only reason it is more important to prohibit killing than to prohibit letting die is that there are more opportunities, and probably more temptations, to kill; second, that simply because it is more important to have a *rule* prohibiting killing than one prohibiting letting die, it does not follow that an individual act of killing is worse than a comparable one of letting die.

But *is* the only reason why prohibiting killing is more important than prohibiting letting die, that there are more opportunities and more temptations to kill than to let die? I think not. Imagine the two worlds as identical in the number and degree of these two kinds of opportunities and temptations. Is not the one in which killing is allowed still morally worse than the one in which letting die is allowed? Surely it is. For one thing, allowing killing gives people *discretion* regarding who dies and when. Such discretion is morally abhorrent in anyone's hands. Permitting people to allow others to die does not give it. For one thing, it would not permit anyone to put people in terminal conditions; and unless they are in terminal conditions, it would be wrong to say without qualification that anyone *lets* them die.

Even if we imagine the two worlds as identical in the number and degree of temptations and opportunities to kill and to let die, there would also be a morally significant difference because of differences in the fears to which people would be subjected. For human psychology being what it is, if killing were allowed, people would tend to fear that someone would kill them. But if killing were prohibited and letting die permitted, the resulting fear would most often be the conditional one that if one had a serious illness or injury, one would be allowed to die.[3] Morally, this seems a less serious consequence for two reasons, both of which would apply even if the relevant fear of being killed were instead that *if* someone wanted to kill one, this would-be murderer might succeed. First, because of what is most often involved in being killed (e.g., someone's performing a gruesome, painful, frightening act), fearing it may be quite horrible and would usually be worse than fearing simply being allowed to die when seriously ill or badly injured. Second, because most people usually suppose that they will not in the near future become so ill or injured as to become candidates for being allowed to die, the fear that if they are in that position they will be allowed to die would be likely to plague them considerably less than would the fear that someone would kill them, were killing permitted.

Now consider the claim that even if, from the moral point of view, it is more important to have a rule prohibiting killing than a rule prohibiting letting die, it does not follow that a particular killing is morally worse than a comparable act of letting die, e.g. an act of letting the same person die under the same conditions, say by unplugging a machine as opposed to giving a fatal injection. This is a difficult claim to assess, but it seems wrong. Certainly it would not follow that any act violating the first rule is worse than any violating the second; and if the idea that letting die is preferable to killing is taken to imply this, then it is mistaken. What we may plausibly hold, I think, is that if doing *A* violates a more important moral rule than doing *B*, then, other things being equal, *A* is morally worse than *B*.[4] One way to phrase the intuitive

---

[3]It might be held that even when killing is allowed the typical fear would be that *if X* wanted to kill one, *X* would succeed. My suggestion is that, in part because one would consider oneself generally liable to being killed (in a way one is liable to being allowed to die only if already ill or injured), one's fear would usually be unconditional. But I shall try to indicate why it would in any case be worse than the counterpart conditional fear of being allowed to die.

[4]In speaking of a more important moral rule I have in mind one that meets the possible worlds test suggested above. At least the main considerations we need to take account of in making the relevant comparisons between possible worlds concern the differences between them with respect to the welfare of the persons they contain. This point is important in preventing the principle from being trivial.

idea here is this: if R is a more important moral rule than R', then R generates more stringent duties, and so, other things being equal, violating R is worse than violating R'.

But what sorts of things need to be equal? All I can do now is suggest some of the relevant variables. One is the condition of whoever is directly wronged by the breaking of the rule. If S is a hemophiliac, then things are not equal with respect to comparing (a) taking a blood sample without S's consent and without a way of stopping the bleeding, and (b) sterilizing a normal male without his consent. For here the sampling would presumably kill S. Another consideration is what *other* moral rules the acts being compared violate. Imagine that someone proposed, as a counterexample to the principle we are considering, that one's not answering one's patient's question about the patient's condition would be worse than lying, provided one knew that not answering would make the patient falsely assume the condition to be cancer and commit suicide. Here we cannot compare just the rule prohibiting lying to that requiring one to answer one's patients' questions about their illnesses; for one is also violating a rule (applicable to doctors in such cases) requiring an effort to prevent patients from committing suicide on assumptions whose falsity one can show them. There are other considerations, but perhaps enough has been said to give some plausibility to the principle that if doing A violates a more important moral rule than doing B, then, other things being equal, A is morally worse than B.

Let me suggest some further grounds for thinking that, other things being equal, if letting S die would be wrong, killing S would be worse. Let us first reflect on the plausible idea that if one lets S die, one need not thereby prevent someone else from saving S.[5] This seems especially plausible if one thinks of killings as more or less instantaneous—which they often are—and of letting die as a slow process. But surely letting die is like killing in entailing death: if Sam is not dead, he may have been *left* to die, and people may be *letting* him die; but it is not true that anyone has let him die until he does die. One may now point out that whereas the acts by which one may let Sam die (e.g., not undertaking surgery) do not preclude others' saving him, the acts

[5]Richard Trammell has made a very similar point to support the moral significance of the distinction between not saving and killing. But as we shall see, not saving does not entail letting die, though the two are sometimes assumed to be equivalent. See "Saving Life and Taking Life," *J. Phil.* **72**, 131–137 (1975). The moral significance of the distinction between killing and letting die is plausibly defended by Daniel Dinello in "On Killing and Letting Die," *Analysis*, **3**, 83–86 (1971), though I believe his characterizations of killing and letting die are both inadequate.

by which one kills do preclude it. This is important when it is so; but it is not always so. Thus, one may kill by poisoning, yet poisoning need not preclude Sam's receiving an antidote.

Not saving, on the other hand, does contrast with killing in that it is consistent with someone else's saving *S.* However, even if intentionally refraining from saving *S* should entail a willingness to let *S* die, it does not entail letting *S* die, since *S* need not die. Thus, we cannot simply assume that arguments for the view that the duty not to kill is more stringent than the duty to save will support a similar thesis about killing as opposed to letting die. One may wonder whether we should not drop talk of letting die and simply speak of not saving as opposed to killing. But these terms all have an irrepressible life of their own, and in any event 'not saving' is quite inappropriate for terminal patients. The trouble is precisely that we cannot save them. One might protest that when we cannot save, we also cannot let die; but at worst we would have to take 'let die' in such cases as short for 'let die sooner than otherwise'.

If we bear in mind the distinction between not saving and letting die, what morally significant contrast remains between letting die and killing, where the fatal act is not one that prevents others from saving? First, killing *S,* unlike letting *S* die, entails causing *S*'s death. Since it is plausible to regard causing a person's death as prima facie wrong, this difference in what killing and letting die entail is reason to take the distinction to have moral significance. Second, killing *S,* in the most merciful ways known, entails invading the body, e.g. with injections, gases, or poisons. Invasions of the body also seem prima facie wrong. Merciful killing may well be possible without doing anything that may plausibly be called invading the body, but the dispatches that come to mind also seem prima facie wrong. These points apply even when the fatal act does not prevent *S* from being saved. Suppose *X* gives Sue an injection that (barring use of an antidote) kills in a few minutes. *X* both causes her death and invades her body. Unplugging Sue's respirator would do neither; and at least partly for that reason it would surely be preferable were other things equal, e.g. if the unplugging were followed by an equally rapid, equally painless death.

It might be replied that letting die is also prima facie wrong. I doubt that; but it is important to note that even if it is, killing *S,* in a situation in which it would be wrong to let *S* die, is objectionable for all the same reasons *and* because it entails both causing death and (if done in the usual ways considered merciful) invading the body. Thus, it would still be prima facie worse, other things remaining equal. This does not entail that mercy killing could never be justified, or even morally preferable to letting die. My aim is rather to show that the weight of evidence favors the view that, other things being equal, letting die is morally preferable to killing. This could be so where other things are equal and (a) both are permissible, or (b) both are impermissible, or (c) letting die is permissible and killing not.

If one imagines letting die as doing something, e.g. unplugging a machine, which is immediately followed by death, it is natural to object that letting S die may cause S's death. After all, X's pulling the plug brings about S's dying at that time. If this means that X brought it about that *it was at that time* that S died, it is true. But bringing it about that it is at t that something happens does not entail bringing about *that* it happens (or causing it), nor is bringing it about that something happens (or causing it) entailed by *hastening* it. Suppose Sam decides that he will commit suicide when he is likely to enter a final coma, and secures a promise from the doctor to tell him the truth when Sam asks whether the time has come. Then the doctor, by telling Sam the truth, brings it about that it is then that Sam commits suicide; but the doctor does not bring it about that Sam commits suicide. Sam has already formed, on his own, the *intention* to commit suicide. He brings the act about; the doctor simply supplys Sam information, thereby determining when Sam carries out this prior intention. On the other hand, if *(i)* 'By pulling the plug then, X brought about S's dying at that time' is taken to mean *(ii)* 'By pulling the plug then, X caused S's death', *(i)* begs the question. In any event, *(ii)* is surely false; for where S dies *of* the illness in the way we are imagining, the illness, not X, causes S's death.[6] If X infected S with it, that is very different; but such cases are quite unlike those in which doctors, nurses, or relatives normally let terminal patients die. In the former, X causes S to have a fatal illness and thereby kills S; in the latter, no person kills the patient.

---

[6]This would *apparently* be denied by Jonathan Bennett, who seems to think that the same act can be both killing S and letting S die. See "Whatever the Consequences," *Analysis* 28, 83–97 (1966). Since I hold that letting S die, without performing any further act, entails *not* causing S's death, I would deny this. If X poisons S and *then* refrains from giving an antidote, X may be said both to have killed and to have let S die; but here we have two different acts. It is not clear to me whether Philippa Foot espouses Bennett's apparent view or not; but she certainly *seems* to think that turning off a terminal patient's respirator may be killing, even in such cases as those I have been arguing represent letting die, but not killing. See "Euthanasia," *Phil. Public Affairs* 6, 85–112 (1977), esp. pp. 101–102. Bennett's example is one of culpably letting S die, and Foot may also be thinking of the act in question as culpable. When letting die is culpable, it may seem natural to call it killing, but this is surely a mistake on balance. Supposing it is a mistake, however, does not prevent us from taking (or require us to take) some acts of letting S die to be as bad as killing S in similar circumstances.

We can now readily see a further point, one which has been overlooked or too little appreciated.[7] The intention to kill differs from the intention to let die; and clearly the difference is such that if X intends to kill S, X has a different attitude toward S's death from that held when the intention is to let S die. First, X in some sense aims at bringing about S's death. If it is prima facie wrong to cause a person's death, then it is plausible to hold that it is also prima facie wrong to aim at bringing it about. Second, on the very reasonable assumption that intending to do something entails being disposed to try to overcome obstacles to doing it, someone who intends to kill S is disposed to try to overcome obstacles in order to bring about that death, even if doing so requires hurting people. Someone who intends to let S die need not be so disposed. If, for instance, X intends to kill S with a drug, X will be disposed to prevent others from giving an antidote (though the disposition may not manifest itself, since, e.g., a colleague X greatly respects might ask to give an antidote); but if X intends to let S die by stopping medication, X need not be disposed to prevent others from (say) trying a new drug. X would be so disposed if the intention were *that S be allowed to die;* but X need not intend this in order to intend to let S die. The assumption that X must intend this may underlie certain conflations of the intention to let die with the intention to kill.

It will help to consider a different kind of example, one that some philosophers have on occasion discussed.[8] Suppose a physician has six patients, five of whom will die unless they receive organ transplants, and one with all the needed organs. Clearly the physician should not kill this one in preference to letting the five die. How are we to explain this except in a way that implies that there is a morally significant difference between killing and letting die? One promising line is to attribute the wrongness of the act to its using a person merely as a means to someone else's end. But this alone will not do. Suppose that five siblings are dying because they are depressed about the apparent death of their mother, and hence will not eat. If all that is needed to save them is the sight of her alive, surely their doctor would be at least morally excusable for forcing her, against her will but without harming her, to go a block out of her way to be glimpsed by the children. Yet she is apparently being used merely as a means to someone else's end.

---

[7]Robert Morison, e.g., says, of what appear to be cases of killing and letting die, "The intent appears to be the same in the two cases, and it is the intent that would seem to be significant." See "Death: Process or Event," *Science* **173** 696–699 (1971). See also the quotation from Fletcher in fn. 2.

[8]See, e.g., Philippa Foot, "The Problem of Abortion and the Doctrine of Double Effect," *Oxford Rev.* **5** 5–15 (1967).

Granted, the doctor would not be justified in doing her serious harm even to save the five lives; but since killing is a great harm (in most people's eyes the greatest) and letting die is, other things being equal, not as great a harm, if one at all, this supports the view that the distinction between killing and letting die allows us to explain, at least in part, why the physician may not kill one to save five. One might also argue that personal property rights to one's organs explain this. But even supposing these organs are one's property, this line will not do. The mother has property rights to her home grown tomatoes; but if giving a few to the five dying children were necessary to save them, taking those tomatoes for this purpose, but against her will, could be at least excusable.

There are other ways one might try to explain why the doctor may not kill one to save five, but I doubt whether any of them will undermine the moral importance of the distinction between killing and letting die. The example is doubly interesting, I might add, because it shows that in at least one important kind of case even several people's right to life, as that right is often understood, cannot override one person's right not to be killed.

Despite all that has been said, (1)–(6) may seem to be undermined by certain examples. A particularly interesting pair is given by Judith Thomson.[9] In one, X "is walking across a field. Unbeknownst to him, a sick baby has burrowed its way under a clump of hay in front of him. He steps on the clump, thereby killing it." The other case is the same except that the clump is alongside X's path and "He walks on; the baby dies; he did not save it." Now although Thomson uses the example to cast doubt on the view that the difference between killing and not saving is morally significant, it may seem to cast doubt on the moral significance of the distinction between killing and letting die. But does X let the baby die? I think not. It is hard to say why not. For one thing, X has no idea that X is in a position to kill or save a baby, nor can we assume here that X should have any idea. For another thing, X's letting S die seems to entail something like this: that X either knowingly refrains from doing something toward S, or negligently or culpably omits doing it. Not saving does not entail anything close to this. (Perhaps it is not saving under some such conditions that those who defend the moral significance of the distinction between killing and not saving have in mind. Thomson is not using the examples, however, to cast doubt on the moral significance of the distinction taken in *that* way.)

In saying what I have, I do not mean to deny that a particular case

---

[9]See "Rights and Deaths," *Phil. Public Affairs* 2 114–127 (1973).

of letting die *could* be worse than a case of killing. Imagine that Sam is a Navy gunner who sees a torpedo about to pass a small ship in a direction in which, if he does not intervene, it will hit and blow up an ammunition ship. He can let its crew of 200 die or blow up the torpedo, thereby killing several people on the nearer ship. It is a comment on the horror of war that, here, not to kill would be the worse choice. A related point can be seen by imagining that Sam lets an innocent person, Ann, die in a painful way, when he can easily save her. Surely this could show him to be as bad a *person* as one who murders her.[10] But evaluating persons is distinct from evaluating their acts, though the two kinds of evaluation are intimately related.

Where does all this leave us with respect to statements (1)–(6)? If we keep in mind that one may let S die even when S is irreversibly comatose (and such parlance is so common that it must be acknowledged, even if we might like to tighten up our terminology), then letting die is presumably not prima facie wrong, nor is there a general right not to be allowed to die. S may have requested to be allowed to die; for this and other reasons, we may be strongly obligated to comply. Thus (1) and (2) seem too strong, though one can plausibly maintain them by allowing the relevant duties and rights to be overridden when it would be morally reasonable to let S die. Statements (3)–(6), however, receive strong support from our discussion. Since even (6)—which is the weakest if it is not equivalent to (5)—would entail that the distinction between killing and letting die has moral significance, I shall consider that thesis prima facie reasonable and apply it to some concrete problems about terminal illness.

II. PATIENTS' RIGHTS REGARDING MEDICAL TREATMENT

A main source of problems in medical ethics is conflicts of rights, e.g. between the right of a patient's spouse, on request, to be given some information about the patient's condition by the doctor, and the patient's right to have certain facts about that condition kept in confidence

---

[10]This seems to me to be what may be shown by an example of James Rachels' *intended* to show that, other things being equal, killing is not worse than letting die. He compares X's drowning a child with X's gleefully letting the child drown, where the child hit its head just before X was to do the drowning, and X stands ready to push the head back under if the child manages to raise it. See "Active and Passive Euthanasia," *New England J. Med.* **292** 78–80 (1975).

by the doctor. I shall be discussing such conflicts in some detail. First, however, we should note that it is very easy to talk about rights without being clear on what their possession entails. To take a very conspicuous example, does our right to life entail just that it is (prima facie?) wrong to kill us, or also that it would be (prima facie?) wrong to let us die? There is no easy way to answer such questions; but it is at least some help to use infinitive constructions (or certain 'that'-clauses) in expressing rights, and I think it best to countenance a distinct right for each distinct state of affairs expressed by such a construction in a true attribution of a right to someone. We may then need to countenance more rights than on a terminology that does not reflect such fine distinctions. However, since some rights can be regarded as the (or a) basis of others, this manner of speaking not only enables us to distinguish rights that we might otherwise confuse, but makes systematizing and interrelating rights easier than it would be if they are as much in need of unpacking as the (apparent) rights to life, liberty, and medical care.

Although rights, like principles or duties, may conflict, some rights not only do not conflict with others, but are the (or part of the) basis of them. For instance, there is surely a right to control one's body and what is done to it. It is largely in virtue of this right that pushing us around physically and the performance of experiments on our bodies without our consent are wrong. If there is a right to control one's body and what is done to it, it is among the most important rights we need to recognize in medical ethics. For on the plausible assumption that all medical treatment in some way affects the body, this right implies another: the right to refuse medical treatment.

Someone might object: We do not have a right to use our bodies to harm others, nor even to resist inoculation when there is a serious epidemic, so how can we have a right to control our bodies? But the first point simply shows that the exercise of a right may be limited by others' possessing that same right; and the second point shows that even the right to control our bodies and what is done to them may be overridden by sufficiently weighty moral considerations, including the fact that a particular exercise of this right would violate the same right of others. If there are any absolute rights—rights which may not be overridden—this is not one of them.

As the foregoing suggests, the right to refuse medical treatment is not absolute either. If S is an adolescent boy who has made a suicide attempt that the physician knows is ill-considered, and S comes into the emergency room comatose, bearing a note by him refusing treatment, the physician would generally be justified in reviving him. Alternatively, suppose that Sue is clearly refusing life-saving treatment either to spite her family, or because of a stubborn misunderstanding which cannot be readily dispelled. If she will die without the treatment, then whatever the legal rule here, from the moral point of view one would hope for a little medical parentalism. On the other hand, adult Christian Scientists have an extremely strong right to refuse even life-saving treatment, though they are still subject to the exception concerning epidemics. Neither our right to refuse to be saved nor even our right to control our bodies and what is done to them gives us a right to refuse to cooperate even minimally in protecting others.

If one believes that there is no right to suicide, one may argue that to regard the right to refuse medical treatment as overridden so rarely as I suggest is in effect to condone suicide. But the distinction between killing and letting die applies to oneself; letting oneself die is *not* suicide. A terminal patient's refusing treatment would typically be at most letting oneself die, or perhaps causing others to let one die. Indeed, even if an accident victim who can be saved refuses treatment and then bleeds to death, this patient does not thereby commit suicide. Such people would prevent others from saving them, but they would die of their injuries, not from an act of suicide.

If there is a right to refuse medical treatment, particularly when one is terminally ill and wants to die, is there also a right to be "put out of one's misery," say, to be given a fatal injection when one is terminally ill and has requested this after careful reflection? If there were, someone would have an obligation to kill the patient in such cases.[11] I do not believe that in every such case someone does; and to count on doctors, nurses, or relatives to do this would be unreasonable. On the other hand, perhaps certain terminally ill adults do have a right to end their own lives quickly. I am inclined to think there is such a right, though surely it can be overridden by a number of considerations, including (sometimes) the likelihood that the patient's family would be very hurt by the suicide.

Even if certain terminal patients have a right to end their own lives, it does not follow that anyone is obligated to try to help them in this, e.g. by supplying drugs. The obligation others would have is chiefly to abstain from trying to stop them from ending their lives. One might try to dissuade, but not to prevent.

---

[11]It is not easy to say how we are to tell *what* obligations on the part of others are implied by S's having a right, and this is one reason why the notion of a right is not as well understood as it may seem. However, among the questions it is helpful to ask are (a) what acts would *violate* the right and (b) under what conditions it is appropriate to assert it.

There is no simple generalization which tells us just what doctors, nurses, or relatives ought to do when patients want to kill themselves. But suppose the physician knows that Sue, a terminal patient, and her family have agreed, upon adequate reflection, that her ending her life is the best course. I would argue that the physician ought to cooperate, while on the case, even to the extent of supplying lethal drugs. Does this imply that even a doctor who deeply disapproved of suicide would have no right to refuse the patient's demand unless this doctor immediately resigned from the case? I think not. There seems to be a weak sense of 'ought' in which one has a right not to do what one ought. Imagine that you have promised to pay me $100 on a certain day, and offer at that time to do so, but explain why it would be quite difficult for you. I have a right to collect, but if I can easily wait the fortnight you suggest, surely I ought to. It would be rather low to insist on my right with no consideration for you, and I would deserve criticism, presumably moral criticism, for this.

Now take a different case. Suppose Sue wants to commit suicide, but the doctor knows that her family would be deeply and lastingly hurt by this, and that death is very close in any case. Here, I suggest that the doctor ought to refuse to cooperate unless an agreement has been made with Sue that specifically requires cooperating; but even apart from such an agreement the physician would have a right to cooperate. On the other hand, where the family believes Sue should end her life and she either is undecided or appears to have agreed under some pressure, the doctor ought *not* to cooperate and ought to discuss alternatives with her. If terminal patients have a right to end their lives, at least it ought not to be exercised unless the impetus comes essentially from them.

So far, I have suggested that except in the special cases where the right to refuse treatment is overridden, normal adults may refuse to be aided by machines (or drugs, therapy, etc.), or even demand the disconnection of machines they initially allowed. From the moral point of view, a valid refusal might consist of suitable directions conveyed before the illness or injury, or in adequately lucid requests after its onset. But the issue is more complicated. For one thing, terminal illness besets not just adults, but children.

May the parents or guardians of children or of the mentally incompetent exercise, by proxy, the *same* rights to refuse treatment which I have attributed to normal adults? Not quite. The rights of parents or guardians to influence S's treatment are less extensive than those S would have if competent to consent. This truth is probably in part grounded on our recognition of the likelihood of biased judgment in at least some such cases. For instance, because the guardians of a retarded, deformed child might be influenced by their desire to unburden themselves, they might prematurely refuse medical treatment to prolong its life, particularly if the physician casts the outlook in a way that maximizes their freedom of judgment.

With children, particularly mentally normal children, there is an added possibility not generally applicable to the elderly terminally ill: the possibility that a miraculous recovery or long-term remission will yield much good life. I am not suggesting that children or the mentally incompetent should never be allowed to die. My point is that they have a need for protection beyond the need created by terminal illness in normal adults, and other things being equal, greater caution and presumably a slower pace are appropriate in allowing them to die.

Another difficult case is raised by terminal conditions which cause a sudden loss of consciousness that leaves the victim in some way unprepared for death. Perhaps Sam has yet to make a will, or the doctor knows that there are important things he meant to say to his family. If there is the possibility of renewed lucid consciousness (roughly, the kind adequate for understanding ordinary conversation), then even if Sam had directed that in a terminal illness or injury he was not to be allowed to "linger" after losing consciousness, it would seem reasonable to try to give him at least one more period of lucid consciousness. Indeed, when a terminal accident, or a fatal illness such as a massive coronary, utterly shocks the victim's family, which is in addition ill-prepared, then even if there is no hope of reviving lucid consciousness, it may be reasonable for the physician to sustain the patient's life for a time despite a directive not to prolong it artificially—unless the directive is so specific on this point as to leave no choice. The doctor might know that a short period of adjustment could greatly relieve the family's suffering. Sam's wife, e.g., might be away on business and his children out of town; they all might want very much not to be away when he dies. Using life-sustaining equipment for even a day or so could fulfill their wishes. It might also be needed to facilitate organ donations, or bodily experiments, to which S has consented. Fortunately, cases of these last two sorts can be dealt with by the patient in advance, but in practice they often are not.

III. PATIENTS' RIGHTS TO INFORMATION ABOUT THEIR CONDITION

There are at least three apparent rights to information we should consider in the patient–doctor relationship, particularly when the patient is terminal. First, surely patients have a right not to be lied to about their conditions. Second, they seem to have a right to have their doctor answer their questions about their condition to the best of the doctor's ability and in a way they will understand. Third, and less obvious, patients might be held to have a right to have their doctor *volunteer* to them the clearest diagnosis and prognosis the doctor can provide, whether they ask for it or not. Let us consider these rights in turn.

First, there surely is a right not to be lied to about one's condition; but I would hasten to add two points. First, this right is not violated by refusals to answer patients' questions about their conditions, nor even by replying in jargon not intelligible to them. Second, this right is surely not absolute; e.g., if a physician knew (however unlikely knowing this sort of thing may be) that S would precipitately commit suicide or would die quite prematurely if forced to face the condition now, and that only lying could avert this, then presumably lying with the intent to offer the truth as soon as possible would be excusable. But the case is artificial. In practice, particularly with the options of changing the subject or having relatives tell S the nature of the condition, the doctor would not be justified in lying to S. The right not to be lied to about one's condition is, I think, seldom if ever (justifiably) overridden.

It is more difficult to explicate the right—and surely there is a right —to have one's doctor answer one's questions to the best of the doctor's ability, and in a manner one can understand. If the doctor believes it will significantly worsen Sue's condition to tell her she will die of it, this may well justify at least putting off answering. It may also justify not answering at all, if, for example, relatives have been told and have agreed to tell her. Granted, in the first case Sue's right is not immediately *accorded* her by the doctor, and in the second, the doctor does not directly accord it at all, but only by proxy. But in neither case should we say her right is violated. Suppose, however, that there are no relatives. Then surely the doctor should answer the questions, if they are repeated, at the earliest reasonable time, even if, because the disease is advancing, there is no opportune time. The problem may be complicated by Sue's having told the doctor either that she does not want her family to know her condition at all, or that she wants to tell it to them herself. Then the doctor needs a stronger justification for delaying an answer to Sue than in the case in which the family knows the prognosis and takes responsibility for informing her.

Concerning the third possibility mentioned, that S has a right to have the doctor volunteer the diagnosis, or at least the prognosis, I do not think there is such a right. For suppose S simply wants not to know, or knows and wants not to hear. Surely the doctor may not then be obligated to tell S, though in some such cases it would doubtless be better to volunteer a prognosis. It may be important that S know there will be no recovery.

On the other hand, given a common kind of doctor–patient relationship, one in which communication is not easy, at least for the patient, Sam's not asking about the outlook is *not* a good reason to think he prefers to remain ignorant of it.[12] And even when it might be best for the doctor not to volunteer the prognosis to Sam, the doctor ought to *provide* an occasion for Sam to ask if he wishes. Indeed, presumably terminal patients do have a right to be provided such an opportunity by their doctors (or a suitably informed, medically trained person, such as a geriatric nurse).

When a terminal illness or injury afflicts a child or someone mentally incompetent, similar problems arise, and here the proxy rights closely parallel those of the normal terminal adult. The parents or guardians have, first, a right not to be lied to; second, a right to have their questions about the condition answered to the best of the doctor's ability in terms they can understand; and third, the right to be provided an occasion to ask about the condition.

There is at least one important respect, however, in which terminal illness in a child or someone mentally incompetent differs from terminal illness in the normal adult: it seems that with respect to the former the physician (or a suitable substitute) *is* obligated to volunteer to the parents or guardians at least a prognosis, even if it seems unwanted. For the welfare of such patients, who are competent neither to waive the right to be given information on request, nor to see their own interests, may well require that their parents or guardians know their fate accurately and promptly. When a child or mentally retarded adult is able to reason, but not able to do so as well as the normal adult, we have an intermediate case where, even more than usual, there will be a need

---

[12]Sometimes, moreover, S's asking for the diagnosis and prognosis is not a good reason to think S *does* indeed want to know them (a point I owe to Dr. Walter Friedlander). S might prefer not to know, but ask out of fear of seeming cowardly. However, if physicians want to appeal to such a possibility to justify not answering or, say, answering in jargon, the burden of showing that the possibility is likely would be squarely on them.

for careful judgment to determine what information to give the patient about the terminal condition and how to give it.[13]

IV. SOME LEGAL IMPLICATIONS

Not every moral principle should be written into law. This applies to moral principles about terminal illness, as elsewhere. But legal principles should at least accord with moral principles and, in general, encourage adherence to moral principles. With this in mind, let us briefly consider some legal implications of the moral rights we have examined.

First, what I have said implies that with at most a few exceptions, physicians who allow their patients to die in accordance with the latter's wishes should not be liable to criminal or civil prosecution, even if there is no living will in force in the situation.[14] The principle which most readily explains this is the simple one that people should not be liable to legal penalties for according others what they have a moral right to. It is in part a recognition of this principle, I take it, that underlies the case for living wills. I shall not try to argue here for or against their legalization; but if what I have said is correct, then we may at least conclude that when such wills are executed, it should be in consultation with at least one's family and physician. Surely, in writing such directives, most people with families have a moral obligation to take some account of their family's (and perhaps friends') needs and wants.

Moreover, it seems desirable that living wills, or other legal instruments for supporting the right to be allowed to die with dignity, be flexible enough to avoid foreclosing options which an unexpected situation may make desirable. For instance, *(1)* it might turn out to be desirable to keep certain comatose patients alive long enough to salvage organs they have donated, even if this means sustaining life longer than

---

[13]In general, the closer Sam is to a normal adult's ability to make competent decisions, the greater the role he should have in medical decisions about him. But it is hard to specify when a developing person acquires the minimal competence necessary to warrant direct consultation, and even harder to specify when "adult competence" is reached. The latter may of course come long before legal adulthood.

[14]Physicians should not have this immunity without adequate evidence that they have acted in accord with S's wishes. From the moral point of view, it is best, but not necessary, that these wishes be in writing and expressed when S is clearly competent. If there is doubt about S's competence at the time, however, protecting S's rights seems to require that the burden of showing incompetence be on whoever alleges it. Senility does not entail the relevant kind of incompetence, which is, roughly, an inability to form and express minimally rational desires regarding one's medical treatment.

otherwise warranted after, say, an accident or coronary. *(2)* The "imminent" death of an unconscious terminal patient might best be delayed in order to provide one more period of lucid consciousness, even if this means using extraordinary procedures. *(3)* A short delay in letting the patient die might greatly reduce the family's suffering (this is particularly likely when unconsciousness occurs suddenly). *(4)* Even a person who wishes to be allowed to die when death is "imminent" may not, on reflection, want life support to be ceased until there is "no significant hope" of renewed lucid consciousness, even if death is imminent (some people—and their families—may hold such oases of clarity very dear). *(5) S* might want to be declared dead on the basis of, say, brain death as opposed to a more liberal criterion, such as cessation of vital functions, or—quite apart from that—want experiments involving the body, for which *S* has given permission, to begin only after cessation of vital functions (this might be an esthetic preference, or might reflect a desire to prevent anyone's declaring death prematurely). *(6) S* may want to avoid very specific language about how high the probability of "imminent" death must be before the doctors declare that there is "no hope," and then carry out *S*'s instructions for termination of treatment; for *S* may wish to leave them free to use their judgment when the relevant probabilities are unclear (*S* may of course still count on a requirement of two or more medical judgments, or on these together with the family's judgment, to protect *S*'s interests).[15] And *(7)*, regarding children and the mentally incompetent, although in certain terminal cases the law should not rule out letting them die, it should in some way reflect both their need for protection and the fact that their parents' or guardians' proxy rights concerning their treatment do not extend as far as normal adults' rights concerning their own treatment.

A crucial question I have left open is whether there is a moral right to suicide, at least for terminal patients. If so, then there should be no legal sanctions against either terminal patients who attempt to end their lives, or against their estates if they succeed, or against their family or doctors if they aid them, provided that the initiative comes from the patient.[16] However, precisely because there is a danger that others may unduly influence terminal patients to end their lives, if no legal sanc-

---

[15]One can see how easy it is to foreclose some of these options, as well as to fail to give legal protection to other options that might be reasonably desired, by studying the Nebraska Natural Death Bill (LB 400), which was modeled on the California counterpart. The former was tabled indefinitely in 1977; the other was passed in California.

[16]The initiative may come from the patient even if the idea is *suggested* by someone else. But coercion and deception, among other things perhaps, need to be ruled out.

tions are to be imposed on complying with their efforts to end their lives, those who aided them should bear the burden of showing that they have not unduly influenced the patients in this. That might require only producing suitable statements from them, but it could well require more.

There is of course much more to be said about the conditions under which terminal patients should be allowed to refuse or terminate medical treatment and about the ways in which they should be treated by their doctors, their nurses, and their families. I have tried to contribute to our understanding of these problems by defending some important distinctions, examining some of the important rights of the terminally ill, and outlining some ways of handling certain conflicts among these rights. I have argued that there is a morally significant distinction between killing and letting die, that there is a right on the part of normal adults to refuse or terminate their medical treatment, and that patients (or their parents or guardians) have a number of rights regarding information about their condition. But having affirmed all these rights, I would emphasize that it is best if doctors and their patients try to structure their relationships so that the *assertion* of rights on either side is unnecessary. Although rights and moral principles are crucial in guiding our practices, one measure of our success in these practices is how infrequently we need to invoke the relevant rights or principles against actual or potential offenders. Moreover, although a number of important principles have emerged from our discussion, it is essential in discussing the doctor–patient relationship, or terminal cases in general, to remember that such principles are not precise, do not cover all the cases that call for moral judgment, and do not always apply in the same way to the cases they do cover. We need them to guide our intuitions; but we must also leave room for individual judgments informed by the facts of the particular case.[17]

[17]This paper has benefited from my discussing one or another topic in it with Dan Brock, Hardy Jones, Eric Kraemer, and Susan Sherwin.

Comments on "The Moral Rights of the Terminally ILL"

Susan Sherwin

Dalhousie University, Halifax, Nova Scotia, Canada

Philosophers, in commenting on one another's papers, often see their task to be the denial of the former speaker's position.* You will be relieved to know that I do not intend to argue that there are no moral rights belonging to the terminally ill. In fact, I am sympathetic with most of Professor Audi's substantive proposals and generally agree with him about which moral considerations must be taken into account. Fortunately for the sake of our discussion, I do disagree with him in some of the arguments offered towards those proposals.

Our most serious disagreement has to do with the question of distinguishing between killing and letting die. Whether or not there is a morally relevant distinction to be drawn between the two is a question that has troubled a good many moral theorists recently. I confess to being confused on the issue. Many examples, including those offered by Professor Audi, seem clearly to reflect a serious difference between the two activities. Others incline one to lean the other way; for instance, consider this set of cases offered by James Rachels in "Active and Passive Euthanasia"[1]

> In the first, Smith stands to gain a large inheritance if anything should happen to his six-year-old cousin. One evening while the child is taking his bath, Smith sneaks into the bathroom and drowns the child, and then arranges things so that it will look like an accident.

---

*The comments that follow are addressed to Professor Audi's paper as originally delivered.

[1]James Rachels, "Active and Passive Euthanasia," *New England J. Med.* **292**, 78–80 (1975).

In the second, Jones also stands to gain if anything should happen to his six-year-old cousin. Like Smith, Jones sneaks in planning to drown the child in his bath. However, just as he enters the bathroom Jones sees the child slip and hit his head, and fall face down in the water. Jones is delighted; he stands by, ready to push the child's head under if it is necessary, but it is not necessary. With only a little thrashing about, the child drowns all by himself, "accidentally", as Jones watches and does nothing.
Now Smith killed the child, whereas Jones "merely" let the child die. That is the only difference between them.

What is required, I think, is an understanding of what underlies our intuitions to see a moral difference in some cases. Here, Professor Audi's discussion provides a clue to the grounds of the distinction which he perceives. He describes two "possible worlds," one in which killing is prohibited and the other in which letting die is outlawed, and comments on how preferable the latter is to the former. With a rather subtle use of rhetoric, he says that "allowing killing would result in peoples' fearing that someone would kill them, permitting people to allow others to die would result only in one's fearing that *if* one becomes ill or injured one *might* not be saved."

What's objectionable, presumably, is that one does not want to die, but prefers to live in either instance, and one's chances are better in the latter world.

But the situation is importantly different in the case of the terminally ill. Some people who are suffering greatly prefer to die than to live. If we allow that death, under certain unfortunate circumstances, is a desirable and acceptable goal for a person (as Audi does), then it is not obviously worse to seek that end by killing than by letting die. Killing is bad because most people choose not to be killed; but if someone quite reasonably wants death, why may it not be granted?

If we are entitled to test moral principles by envisioning possible worlds, then for the purpose of this discussion we should consider a world in which both killing and letting die are prohibited except for the hopeless case of an individual who is suffering greatly from an incurable terminal condition. Is such a world, which permits quick, painless killing of those who are suffering incurably and who choose to be killed, necessarily worse than one that only permits letting them die (sometimes a rather lengthy process)? It seems to me that the relevant question here is how we go about determining the criteria for voluntarily terminating the life of a person. Professor Audi's examples, like those of most philosophers engaged in this dispute, are concerned with choosing the lesser of the two evils, when death is viewed as a bad outcome in either event. Such considerations do not seem relevant in situations where we perceive death to be a blessing.

Furthermore, his criterion of a difference in intention is also not applicable in these cases, for in hastening the end of a terminally ill patient, the intent to kill is not different from an intention to let die. Nor is there any distinction to be drawn, as he suggests, on the basis of discretionary powers. In the case of the terminally ill, our choice is one of prolonging but not saving life. We cannot save and heal those patients, only maintain their lives a bit longer. But since a choice not to prolong life is still a choice, the question of discretion comes into play for choosing to let such patients die as much as for choosing to kill them.

There is another matter I find troublesome in Professor Audi's approach: I am uncomfortable about discussions couched in the language of "rights." The term "rights" is used to refer to a wide variety of claims, not all on a moral par. Rather than illuminating this murky terrain as he hoped to do, I fear that Professor Audi has gotten rather bogged down in it.

Consider his argument on the duty of the medical staff to a patient seeking euthanasia. He argues that although there is a right to refuse medical treatment, there cannot be a right to be "put out of one's misery," because such a right would entail that someone has an obligation to kill the patient. However, this does not follow immediately, for it is not the case that all rights have such correlative duties. I have a right to marry, but no individual has the obligation to marry me; and I have the right to vote for the candidate of my choice, but the candidate of my choice has no duty to run. So I might have the right to be killed without anyone having the duty to kill me. Nonetheless, I think this case is one in which the patient's right, if it exists, does entail that someone might have the duty to do the unpleasant act. Audi mentions supplying drugs as one possible means of arranging the patient's death. The situation we have is that the medical profession has reserved the distribution of drugs as their exclusive domain: they have made it illegal for anyone else to provide most drugs. Thus, if someone is entitled to drugs, I should think members of the medical profession do have an obligation to provide the drugs precisely because they have made it impossible for the individual to get what is needed elsewhere. If there is a right to die and if another person's medical cooperation is required to exercise this right, then the right to die is probably a right like the right to education (and unlike the right to marry) in that it does entail specific duties on the part of individuals in certain relevant roles.

Professor Audi argues that his organ transplant example "shows that in at least one important kind of case even several people's right to life, as that right is often understood, cannot override one person's right not to be killed." I'm not sure what the right to life requires of us (if there is in fact such a right). But in the text following, Audi does seem to recognize that under some circumstances several people's "right to life" does outweigh someone's "right not to be killed": "Killing one person might be required to save many people whom he is threatening." There is no explanation for this inconsistency. Moreover, it has been persuasively argued[2] that a person's right to life (and right not to be killed) may be overridden not only by another person's right not to be killed, but even by a single person's simpler rights to autonomy.

If Professor Audi is comfortable in using the concept of "rights," I would point out to him that in Section II, he argues for a fundamental right to control one's body and what is done with it. The existence of such a right is sufficient to account for his intuitions about not killing one patient for the organs to save five others, and hence undermines one more ground for postulating the dubious killing–letting die distinction. In fact, this right seems sufficient to generate all the normative conclusions he is pursuing. I wonder that he does not have it do more work, except perhaps for some shared discomfort about how hard it is to establish that we do have such a right, and uncertainty where it fits in the overall scheme of rights. (Which other rights override it, and which does it override?)

I also have three quibbles which are rather more minor, so I shall just mention them briefly. In the final section, Professor Audi argues that people should not be liable to legal penalties for doing what is morally permissible. Most of our laws are, in fact, concerned with making people liable for doing things that are morally permissible— driving above 100 kilometers per hour on a clear highway on a clear day, operating a restaurant without a license, educating one's own children. Perhaps the relevant point is that it is morally permissible for physicians to allow their patients to die, and that there are no further reasons for making such action illegal; and therefore it ought not to be subject to legal penalty. On the basis of Professor Audi's argument, the case seems even stronger: it is sometimes morally required to allow patients to die, and the law certainly ought not to prohibit individuals from fulfilling their moral duties.

I am also suspicious of the principle "if doing A violates a more

---

[2]See Judith Jarvis Thomson, "A Defense of Abortion," *Phil. Public Affairs* 1, 47–66 (1971).

important moral rule than doing *B,* then, other things being equal, *A* is morally worse than *B.* " This statement seems either to be false or tautologous, depending on the work to be done by the "other things being equal" clause.

And, finally, my intuitions run in rather the other direction in the case of the person "who has made a suicide attempt that the physician knows is ill-considered [and who] comes into the emergency room comatose and clutching a note refusing treatment." Here, Professor Audi remarks, "the physician would generally be justified in reviving him." My inclination is to doubt that the physician *knows* that the attempt was ill-considered; generally, we ought to bear this condition of uncertainty in mind if we are genuinely to respect individual control over one's own body.

II

ISSUES IN GENETICS

5

On Getting "Genetic" Out of "Genetic Disease"[1]

Richard T. Hull

State University of New York, Buffalo, New York

I. INTRODUCTION

Most of the attention devoted in recent years to the ethical issues involved in the manipulation of human genes has been focused either on the potential risks involved in acquiring the knowledge needed to manipulate genes effectively, or on the valuational questions associated with the use of positive eugenics to improve the species. In illustration of the former sort of issue, the highly touted Asilomar Conference endorsed limitation of research on transplantation of genes into viruses and bacteria to strains unable to survive outside of highly artificial laboratory environments; the restriction was adopted in light of the potential risk of deadly mutant strains escaping the confines of the laboratory.[2] Concerns of the latter sort, ignited by futurists such as H. G. Wells and Aldous Huxley, and fueled by the studies of Arthur Jensen and others into the genetic connections between intelligence, race, and socioeconomic status (together with the opposition to and criticism of their work), have prompted reassurances from such distinguished researchers as Bernard Davis, who has argued that our "vast

---

[1]An earlier version of this paper appears under the title "Why *'Genetic* Disease'?" in Marc Lappé, Tabitha Powledge, *et al.,* eds., *Genetic Counseling: Facts, Values and Norms,* New York, Stratton Intercontinental Medical Book Corporation, forthcoming. Material from that version is herein reprinted with the kind permission of the editors and publisher. The research for these papers was sponsored in part by the Institute of Society, Ethics and the Life Sciences, Hastings Center, Hastings-on-Hudson, New York.

[2]"Government Report," *The Sciences* **15**, 29 (1975).

ignorance" of the genetic and environmental bases of polygenetic traits "protects us against the main possibilities of harm from gene replacements."[3] By contrast, most commentators do not question the inherent value of using techniques of genetic engineering to "replace defective genes," "cure disease," and encourage expression of "single genes that influence many aspects of man's health (such as specific immune responses)".[4] Indeed, these terms, and associated ones, are generally treated as though their application is a value-free matter, involving no intrinsic appeal to anything but objective fact.

Interestingly enough, however, it is from the arena of public policy that this attitude toward disease terminology has recently been questioned. In an address given at the California Institute of Technology in 1972, Senator John V. Tunney asked, "(H)ow are words such as 'normal,' 'abnormal,' 'health,' 'disease,' and 'improvement' defined? Are they words that can be operationally used to determine what should be done in the area of genetic engineering? . . . Questions such as these . . . can be answered only by appealing to ethical, or so-called moral, arguments."[5] Philosophers these days customarily busy themselves with the analysis of language and concepts, seeking to uncover and examine presuppositions on which other disciplines turn. Thus philosophers, just as do politicians, claim a license to tamper in others' business. Being a philosopher, I hold such an operator's license. So, perhaps it won't be regarded as too presumptuous if I propose an analysis of a range of genetic terminology frequently employed by geneticists, genetic counselors (including physicians), and even politicians. This analysis seeks to expose a strong valuational component in that range of terminology, and that exposition will in turn form the basis for my argument that most usages of that range of terminology should be strongly curtailed, if not avoided entirely.

* * * * *

There is a cluster of words and phrases which are continually employed in genetic counseling, therapy, and research, but whose careful analysis has been generally neglected. The more obvious of these (with the explicit valuational component italicized) include expressions like "*mal*adaptive trait," "congenital *mal*formation," "*defect,*"

---

[3]Bernard Davis, "Genetic Engineering: How Great is the Danger?", *Science* **196**, 309 (1974).

[4]*Ibid.*

[5]John V. Tunney and Meldon E. Levine, "Genetic Engineering," *Saturday Review* **55** (32), 23–28 (1972).

"deleterious gene," and "inborn *error* of metabolism." Even such terms as "genetic abnormality" and "chromosomal aberration," indicative of only a deviation from the normal range of variability, are most frequently used in application to disvalued divergence.

Less obviously, there are even valuational components to the meaning of the important and constantly-used expression "genetic disease," and I want to examine them in this paper. These components are in part etymological, stemming from the origins of the term "disease." But they also derive from two other sources: the social and psychological results of identifying a disease with only one of its causal conditions, and the consequences of the selective focus on one particular range of possible therapies, prevention, and research stemming from the decision to call a disease "genetic." In the following sections, I will trace out these sources of the valuational aspects of the expression "genetic disease," to show the ways in which they can come into conflict in the arenas of genetic counseling, therapy, research, and even legislation.

II. ETYMOLOGY OF "GENETIC DISEASE": SOME SHIFTS IN MEANING

A. "Disease"

Conceptually, the notion of disease originates in the patient's complaints about bodily discomforts. A disease is identified, first and foremost, in terms of the dis-ease involved. In the subjective sense of the term we speak of symptoms, or abnormal feelings which the patient has. But a disease is also identified in terms of the abnormal states or conditions of the body that are publicly observable; these are sometimes also called symptoms, but I shall refer to them as signs. Some signs may be so obvious that they become part of the patient's complaint (such as rashes, paralyses, etc.), but others may require more skilled observation to detect (such as dullness in some portion of the lungs under percussion).

Apart from physiological knowledge of how an individual symptom or sign arises and may be made to recede, often little else is required for mere palliation. But the notion of disease has also come to involve, first, the idea of a syndrome, and, second, the idea of its cause. A syndrome is a set of *associated* symptoms and signs, associated both through their frequent joint occurrence and through the notion that they have a common cause. For we are frequently not satisfied with mere palliative measures: they must be constantly administered throughout the disease's natural course to suppress the occurrence of signs or symptoms, and are often not fully successful. They involve conceiving of and treating symptoms individually, whereas we suspect that in most cases constellations of signs or symptoms can be thought of as manifestations of a common cause. Hence, any effort to think of dis-eases causally, with an eye to their avoidance or cure, leads to the notion of a syndrome—a constellation of signs or symptoms thought to have a common cause.

But it is rather overly simple to speak of "the cause" of a disease. One of the benefits of philosophers' investigations into the term is to show how idiosyncratic our use of "cause" can be. In its fullest sense, of course, the cause of something is the total set of its individually necessary and jointly sufficient conditions. In talking of the cause of a disease in more ordinary circumstances, however, this definition of "cause" is not employed, in large part because of the difficulty of satisfying its demands. Rather, "some one or more conditions within that set which were novel, unusual, or controllable" are usually cited in giving the cause.[6] Professor Michael Scriven once described this common notion as follows: "A cause is a non-redundant member of a set of conditions jointly sufficient for the effect . . ., the choice between the several candidates that usually meet this requirement being based on considerations of context."[7] For, what strikes us as novel, unusual or controllable will be singled out as the cause just as our interests lie in uncommon, novel or controllable phenomena.

Consider the following illustration, which is an example I recall from Professor Scriven's lectures. A college co-ed has a premarital affair with a male student of whom her parents strongly disapprove and who has promised to marry her, and she conceives as a result. When he learns of it, her boyfriend jilts her. She comes from a conservative home and does not feel that she can turn to her parents for help and advice. Her roommate is studying hard for exams and rebuffs her attempts at soliciting a sympathetic ear. She becomes even further depressed over the reaction of the infirmary physician, who confirms the pregnancy and lectures her on her loose ways, but offers nothing but scorn at her tentative broaching of the subject of an abortion. She

---

[6]Richard Taylor, "Causation," in Paul Edwards, ed., *The Encyclopedia of Philosophy,* Vol. I, New York: Macmillan and Free Press, 1967.

[7]Michael Scriven, "Explanations, Predictions, and Laws," in Herbert Feigl and Grover Maxwell, eds., *Minnesota Studies in the Philosophy of Science,* Vol. III, Minneapolis: University of Minnesota Press, 1962, p. 215.

wanders out onto the San Francisco Bay Bridge, and stands looking over the rail. A passing motorist yells "Jump!" She climbs onto the rail; other motorists stop and watch her. She leaps off and drops into the waters of the bay. What was the cause of her death? Any of several of the conditions surrounding her could plausibly be pointed to as the cause (in the sense of "cause" that Scriven describes), in that we can point to a number of necessary conditions in the sequence of events that led up to her death. Such are events which, had they not happened or been prevented from happening, would have rendered the set of prior conditions insufficient to bring about the effect; depending on where our interests lie, we will cite some one or more of these as causes of her death. The coroner may list drowning as the cause; a policeman may list the cause as suicide by jumping off the bridge; various sociologists and psychologists with varying research interests may point to the seduction and abandonment, the coldness of the physician, the alienation from the parents, the preoccupation of the roommate, or even the jeers and passivity of the drivers on the bridge as the event or events responsible for her death. The idea of cause as we most commonly employ it embodies the concept of human interest, in that it is in large part this that will determine which of several possible candidates for the cause of an event we will so identify.

The notion of a syndrome—a set of associated signs and symptoms thought to have a common cause—is not quite sufficient to capture the concept of disease as it is most commonly employed today. Indeed, there are some dramatic departures from the notion of disease as rooted in the patient's characteristic complaints together with associated signs. In its efforts to understand, control, and avoid disease, modern medicine has incorporated into the very identification of a disease the notion of the cause of the syndrome. This permits the individuation of similar syndromes with distinct causes into different diseases. For example, the general classification of epilepsies along syndromic lines (petit mal, gran mal, temporal lobe) is augmented by classifications amounting to discrimination by cause; each type is further divided into ischemic, traumatic, and ideopathic. Thus, there are now three kinds of petit mal epilepsy; the symptoms are no longer definitive of the disease, but constitute manifestations of the disease; the disease is now identified with the underlying cause of the symptoms; and ischemic and traumatic gran mal epilepsy are spoken of as distinct diseases which manifest similar symptoms.[8] What has happened is that the root notion of disease

[8]J. Pincus and G. Tucker, *Behavioral Neurology,* New York: Oxford University Press, 1974.

—the syndrome—is replaced in its identification by the syndrome's cause. I shall return to this significant point.

B. "Genetic"

There are two things to be said about the "genetic" component of the expression "genetic disease." The first can be put in a couple of ways. Every nongenetic disease has a genetic component, describable in terms of the presence (or absence) of one or more genetically-determined traits. Susceptibility or resistance to infection by a given microorganism, injurability, and toxic reactability are each in part a function of genetic endowment (although doubtless more frequently polygenic than monogenic in character). We may well suppose that individual resistance, say, to smallpox, is a function of genetic endowment, yet we do not classify smallpox as a genetic disease. Why? Another way of putting the same point is this: every nongenetic disease has both genetic and nongenetic members of its set of nonredundant, jointly-sufficient conditions. This raises the question of why and how some diseases come to be classified by appeal to their genetic component(s) and others by appeal to their nongenetic component(s). Whence our interest in only certain members of those sets of jointly sufficient conditions, and not others? I shall return to these points as well.

The second observation is this. Many so-called genetic diseases have known environmental components; the associated syndrome is manifested only in the presence of specific environmental factors or agents. For example, the hemolytic anemia associated with glucose-6-phosphate dehydrogenase deficiency (called "favism") is manifested only upon exposure to certain foods (such as *Vicia faba,* the common broad bean) or certain drugs (such as primaquine, used in treatment of malaria).[9] Xeroderma pigmentosum skin cancer is triggered by sunlight and ultraviolet radiation, but otherwise would not occur in persons homozygous for the recessive gene. Further, many genetic diseases involve the deficiency, or deficiency at a crucial stage of development, of some enzyme or hormone that plays a developmental or maintenance role. Insulin, whose absence produces the hereditary metabolic disease diabetes mellitus, can be artificially supplied, thereby allowing the patient to live a relatively normal life. Homozygotes for phenylketonuria

---

[9] R. King, *A Dictionary of Genetics,* 2nd ed., New York; Oxford University Press, 1974.

are unable to convert phenylalanine to tyrosine because of the lack of the liver enzyme phenylalanine hydroxylase.[10] The weight of evidence indicates that the presence of phenylalanine and its metabolites in large quantities hinders the laying down of myelin. Hence, a low phenylalanine diet, instigated at birth, allows more or less normal development of myelin to occur, thus avoiding or minimizing the associated mental deficiency. The diet may be relaxed after the first several years, since most myelin is laid down then and only maintenance of the sheaths is carried out thereafter. The point is that in these cases the absence of the enzyme is "the cause" of the disease; for the enzyme could (in principle) be supplied artificially, or the results of incomplete metabolism could be avoided by dietary control. Sodium cyanate has been used clinically in the treatment of sickle-cell anemia (although it is a suspected toxicant of the central nervous system); and a recent report indicates that dimethyl adipimidate may act on hemoglobin to increase its affinity for oxygen and restore the red blood cell membrane to normalcy without affecting red cell metabolism.[11]

Thus even diseases such as sickle-cell anemia, which seem under normal conditions to possess causally-sufficient genetic antecedents, frequently are manifested through intermediate steps that are capable of reversal or interruption.

Finally, a number of genetic diseases that are thought to owe to a single (dominant or recessive) maladaptive allele may in fact involve the absence of some offsetting additional mutation. For example, several cases have been reported of healthy adults in families having Tay-Sachs or Sandhoff's disease in which the healthy adults appear to have a double dose of the associated lethal recessive gene.[12] The evidence suggests that the effect owes to the presence of an offsetting mutation in the gene. Thus, if the presence of the second mutant is a sufficient condition for the nonoccurrence of Tay-Sachs or Sandhoff's disease, its absence is as much a necessary condition for each of those diseases as is the commonly associated so-called lethal recessive in double dosage. What at first appears to be a classic case of the presence of a genetic

---

[10]*Ibid.*

[11]B. Lubin, *et al.*, "Proceedings of the National Academy of Sciences," *Science News,* 107 (1975), p. 136.

[12]R. Navon, B. Padeh, and A. Adam, "Apparent Deficiency of Hexosaminidase A in Healthy Members of a Family with Tay-Sachs Disease," *Amer. J. Human Genetics* 25, 287–293 (1973); J. Vigdoff, N. Buist, and J. O'Brien, "Absence of β-N-Acetyl-D-hexosaminidase A Activity in a Healthy Woman," *Amer. J. Human Genetics* 25, 372–381 (1973); J.-C. Dreyfus, L. Poenaru, and L. Svennerholm, "Absence of Hexosaminidase A and B in a Normal Adult," *New England J. Med.* 292, 61–63 (1975).

factor homozygously fitting our definition of cause, so that we could without embarrassment identify it as *the* cause of the disease, turns out not to be so unequivocally.

This latter sort of case suggests a way to view even the most genetic of diseases in a nongenetic way. Our general understanding of such disorders indicates that in them the presence or absence of certain gene products in the internal biochemical environment of the body are the causes of those genetic diseases with no identifiable external environmental factor. But this raises the theoretical possibility of modifying the internal environment by means of introducing the essential product at its crucial point of interaction in some metabolic chain, or of suppressing the production, or otherwise offsetting the effect, of some gene product. Thus, strictly speaking, the presence or absence of the gene itself is not the cause of the disease, but rather the presence or absence of one or more chemicals which serve as partial determinants of the inner environment. To the extent that such products theoretically can be suppressed or supplemented by externally-originating action, we may fairly characterize even the most "genetic" disease as having non-genetic components.

These cases suggest the possibility that no disease (or at most very few diseases) has only genetic factors involved in its total set of individually necessary, jointly-sufficient conditions. It is evident that in most cases of genetic disease the term's application reflects a choice having been made among the causal factors to emphasize the genetic component and deemphasize the environmental component. And there is some inductive reason to suppose that even the hard core of what now seem truly "genetic" diseases may shrink as our knowledge of euthenic engineering increases, and we become able to control expression of the existing genetic information of individuals affected with diseases having genetic components, so as to lead to more desirable phenotypes.[13] Hence, I am tempted to endorse the generalization that every "genetic disease" has non-genetic components, whether presently known and understood or not.

If I am correct, my argument has so far established the points that, in general, identifying a disease involves citing some factor as the cause of a set of symptoms and signs, and that that citation involves selecting one among a variety of candidates in accordance with certain interests. I also claim to have made a strong case for regarding these general truths as holding in the case of so-called genetic diseases. What is

---

[13]Joshua Lederberg, "Biological Future of Man," in G. Wolstenholme, ed., *Man and His Future,* Boston; Little, Brown, 1963.

needed at this point is some insight into the way interests determine the identification of a disease as genetic.

III. CRITERIA FOR APPLYING "GENETIC DISEASE": SOME VALUATIONAL DIMENSIONS

Sedgewick[14] and others (e.g., Dubos[15]) have defended the view that "illness and disease, health and treatment, . . . [are] social constructions. . . . All departments of nature below the level of mankind are exempt both from disease and from treatment—until man intervenes with his own human classifications of disease and treatment." Sedgewick suggests a set of criteria for something being a disease, which I here modify so as to be necessary and sufficient for applying the term "genetic disease." For one to speak of something being a genetic disease, there must be reference made to: (1) a species in the welfare of whose members one takes an interest; (2) a state or item of behavior of some relatively small number of those members that deviates significantly from what one recognizes as norms of appearance and behavior; (3) states or items of behavior that one disvalues; (4) only such states and items of behavior for which family history provides evidence of heritability, or for which our present scientific theories and technologies provide an identifiable genetic causal component; (5) only such states and items of behavior for which the present genetic technology and clinical practice hold promise of alleviation or avoidance (even if only by identification and nonreproduction of carriers); and (6) only such states and items of behavior for which alternative, nongenetic approaches to treatment or avoidance one disvalues.

Professor Michael Ruse has pointed out to me that we tend to classify schizophrenia as a genetic disease, but to prefer to treat it with drugs, and that this serves as a counterexample to criterion (6). I should think, however, that treatment with drugs is not an alternative to some other mode of treatment involving a genetic approach; it *is* the genetic mode of treatment standing as an alternative to other nongenetic approaches such as psychotherapy. To see this we need to consider just how the operations of drugs such as chlorpromazine are thought to occur. The current genetic models of schizophrenia premise that the symptoms of the disease owe to the presence of either too much dopa-

[14]P. Sedgewick, "Illness—Mental and Otherwise," *Hastings Center Studies* **1**, 19–40 (1973).

[15]René Dubos, *The Mirage of Health,* New York, Harper & Row, 1971.

mine in the mesolimbic dopamine tract or too many dopamine receptors; in turn, any increased amount of dopamine arises either from an
increase in the absolute quantity of dopamine produced, or from a
decrease in some offsetting enzyme that normally transforms excess
dopamine. Whatever the deficiency or excess, it is in turn the result of
a genetic variant exhibiting heritability in identical twins of from 40 to
80%, less in fraternal twins, less in more distant relatives, and an even
lower concordance in unrelated members of households. Chlorpromazine blocks dopamine receptors, thereby reducing the sensitivity of the
mesolimbic dopamine system to the increased quantity of dopamine.[16]
Since this form of treatment is preferred over psychotherapy by those
who, like Professor Ruse, classify schizophrenia as a genetic disease,
and since it operates under a model that I would want to call part of
the genetic technology (including not only supplementation of missing
gene products but also compensation for an overabundance of them),
this seems to me to stand as a confirmation of my criterion and not as
a counterexample.

   Several other features of these criteria warrant specific comment.
First, it should be borne in mind that they are intended to be descrip-
tive. That is, they purport to describe the conditions that obtain when
someone calls a malady a genetic disease. They are not intended to be
prescriptive of how or when the term ought to be employed—all the
more so since it is one of the theses of this paper that the term ought
to be eliminated. Second, note that it is a feature of this analysis that
the term "genetic disease" has been relativized in a number of ways: to
species of interest, to recognized norms, to individual and societal
values, to current scientific theory, and to current and forseeable technology. In turn, each of these is either value-laden or in some way
determined by values. Given that pervasive interlacing of valuational
considerations throughout every aspect of our operative criteria for
"genetic disease" (and "genetic health," if its definition contains "the
absence of genetic disease" as a necessary component[17]), we must conclude that the term is valuational both in its implications (e.g., that
alternative typologies have been discounted) and its applications. A
similar sort of analysis can be provided for the other terms mentioned
at the start of this essay. Third, the criteria do not dictate what is to
be called a disease; they only describe the conditions that are generally

---

[16]Tyrone Lee and Philip Seemen, "Dopamine Receptors in Normal and Schizophrenic Brains," *Soc. Neuroscience Abstr.* **3**, 443 (1977).
   [17]Marc Lappé, "Reflections on the Cost of Doing Science," *Ann. New York Acad.
Sci.* **265**, 102–111 (1976).

satisfied when one calls something a genetic disease. Specifically, either a syndrome or its cause could be labeled a genetic disease. I shall show how that choice of label may itself have important moral consequences.

Putting together the results of our enquiry, we see that the decision to call a disease "genetic" involves its identification with a non-redundant or individually necessary member of the set of conditions jointly sufficient for the associated syndrome, where there may be alternative necessary conditions with which it is not identified owing to our relative lack of interest in them; further, the decision to call a disease "genetic" involves shifting the application of the term "disease" from the syndrome to "the cause," and the designation of the syndrome as a *manifestation* of the disease.

As Sheldon has observed, "the basis of the categorization of disease fundamentally determines what gets done about disease, even more than the organizational structures which develop to deal with it."[18] If so, the decision to identify a disease as genetic may well determine a preference of a particular range of possible foci for therapy, prevention, and research. It thus becomes legitimate to ask, What are the moral dimensions of this "decision"?

IV. USES OF "GENETIC DISEASE": SOME POLICY IMPLICATIONS

The positive benefits are by now obvious and familiar. They involve (1) furthering our knowledge of human development and variation as determined by the fundamental processes of human reproduction; *(2)* demythologizing the occurrence of various kinds of disease (in particular, providing more cogent explanations of various disorders than their supposed visitation on persons as punishment for sin); *(3)* the possibility of eradicating various kinds of disease by elimination of their genetic determinants from the gene pool; *(4)* the possibility of improving the species and thus exerting control over evolutionary processes; and so forth. Rationales of these sorts generally hold up select long-term consequences in justification of the pursuit of genetic research, expressed in terms of the elimination of certain diseases, the advancement of knowledge, perhaps even the enhancement of desirable characteristics of the species, as likely future benefits.

But of course all such goals can be pursued whether we em-

---

[18]A. Sheldon, "Toward a General Theory of Disease and Medical Care," *Science, Medicine and Man,* 1, 237–262 (1973).

ploy the typology "genetic disease" or not. Given that employing
the term diagnostically does contribute in some way to a social and
policy-making climate favorable to those goals, whether they provide adequate justification for such a use depends on whether there
are significant negative considerations relevant to their employment.
(I ignore, for present considerations, the question of whether even
these goals may not have negative aspects.) The following considerations are divided into two classes: those which turn on the identification of *any* disease in causal terms, and those which involve the
genetic typology.[19]

Neville has observed that ". . . short of a complete understanding
of all the causes of a disease, decisions as to which to investigate are
contingent upon the idiosyncracies of the clinicians and investigators,
the state of their arts, and the policies channelling research and treatment funds."[20] Few, if any, diseases are so well understood that we can
justifiably identify the total cause and define the disease in terms of it.
Yet, as Engel[21] notes, there is a profound tendency in humans to think
in terms of substantive, unit causes. He cites two examples: the germ
theory of disease, postulating as it does the existence of invisible, omni-
present entities which produce disease, had such a wide appeal that it
became a dominant theme in medicine for some time, causing the
earlier emphasis on homeostasis and other factors within the host to be
relatively discounted. More recently, with the development of pathol-
ogy and diagnostic techniques to detect functional and structural devia-
tions, "the temptation has been to consider these findings the explanations for a disease rather than manifestations of the disease state. A
disease, then, has substantive qualities, and the patient can be cured if
the diseased ('bad') part is removed." To take the successes of surgical
excision in treating disease as evidence for internal irregularities being
the causes of the diseases is fallacious, since it ignores the role of other
necessary conditions as well.

To build one causal factor into the very conception of a disease and
thereby to prefer one causal hypothesis over others is to invite stagna-
tion of research and treatment, and distortion of funding priorities. A
case in point is that of the viral conception of the etiology of cancer,
which has held sway in the funding of cancer research to the detriment

---

[19]Cf. Lappé, *op. cit.*

[20]Robert Neville, "Gene Therapy and the Ethics of Genetic Therapeutics," *Ann.
New York Acad. Sci.* **265**, 153–161 (1976).

[21]G. Engel, "A Unified Concept of Health and Disease," *Perspect. Biol. Med.* **3**,
459–485 (1960).

of other approaches that have a higher likelihood of payoff in increased rates of cure.[22]

Settling upon causal specifications discourages alternative conceptions of disease. For example, it has been recently observed that "The success of Western science and related factors have produced a form of ethnocentrism, with biomedical diseases seen as the only real ones. This may have had the effect of rendering the search for new paradigms about disease partially illogical and inappropriate."[23] The classic Western conception of disease focuses on its biological import as specified in terms of disruption of normal functioning of biological subsystems of the body. But an ethnomedical approach to disease would focus on "behavioral paradigms of disease . . . as devices for codifying and measuring a person's social functioning [where] . . . social behavior correlates of all kinds of interruptions in functioning are delineated regardless of the individual's culture. It is irrelevant (to an ethnomedical approach) whether outsiders judge that the alterations in behavior are caused by changes in sugar metabolism, toxic effects of an infectious or neoplastic disease, anxiety and depression, or, for that matter, the effects of preternatural influences. What is relevant, however, is the time-related changes in which the form of social functioning is altered or interfered with and/or the changes in the way the person uses social symbols."[24] A theory-neutral conception of disease in terms of symptoms and signs, although not independent of subjective factors, does permit the development of such alternative paradigms of disease as behavioral and biological ones which selectively emphasize complementary aspects of the disease process. But to conceive and define a disease in terms of one causal aspect is to impose a priori the appearance of implausibility and confusion on other approaches.

Within the context of our own cultural institutions and practices in dealing with disease, some more specific negative consequences of the typology of "genetic disease" can be seen. Both Sheldon[25] and Engel[26] have emphasized the effect of the nomenclature employed in identifying a disease on therapeutic and preventative measures elected by the physician. "While our understanding of the processes involved is constantly changing, names are rarely changed, and they often continue to exert a weighty influence on the physician's concepts, at times blocking any

[22]A. Kopkind, "The Politics of Cancer," *The Real Paper,* March 31, 1976.
[23]Engel, *op. cit.*
[24]*Ibid.*
[25]Sheldon, *op. cit.*
[26]Engel, *op. cit.*

research contrary to what the name implies."[27] In this way, identifying a disease with one range of necessary conditions (say, genetic) focuses attention on one range of treatment and/or prevention. This in turn issues in perceptions of false or inappropriate dichotomies; for example, either one aborts a fetus identified through amniocentesis and karyotyping as having a genetic disease, or one produces an offspring with the disease. This dichotomy is false in cases where there exists an alternative way of avoiding the symptoms of the disease. For example, it is possible to diagnose glucose-6-phosphate dehydrogenase deficiency prenatally. However, the X-linked enzymatic abnormality does not automatically result in the associated hemolytic anemia, but only upon dietary exposure to certain foods and drugs. Avoidance of those foods and drugs is an alternative to both therapeutic abortion and production of an offspring with the disease symptoms, but this alternative may be obscured as a way to avoid the disease by identifying the disease with the genetic antecedent. With genetic diseases for which prenatal diagnosis is not possible, such as phenylketonuria and galactosemia, avoidance of production of an individual afflicted with the disease would be possible only if reproduction were denied altogether. But again this is a false dichotomy; in the case of PKU strict observance of a phenylalanine-free diet for a dozen years or so avoids the associated mental retardation, and a similar result is achieved with a galactose-free diet in cases of galactosemia. Even in writers who appear not to fall trap to such false dichotomies, one finds evidence of a conceptual distinction between a disease and a set of symptoms as its manifestation that indicates that the disease has been identified with the cause. For example, in discussing the use of phenylalanine- and galactose-free diets to avoid the mental retardation associated with phenylketonuria and galactosemia, Murray writes: "The dream of the physician is, after all, the prevention of *the manifestation of disease* rather than its cure after it has been found"[28] (italics mine). Put this way, it is the manifestation of the disease and not the disease itself that is prevented in the case of genetic diseases with manipulable components; possession of the genetic component is a sufficient condition for possession of the disease, although not for its manifestation. Under this sort of conception, the genetic disease can be prevented only by preventing individuals who possess the genetic component.

The foregoing point is closely related to another one, namely, the

---

[27] *Ibid.*

[28] Robert Murray, "Screening: A Practitioner's View," in Bruce Hilton *et al.*, eds., *Ethical Issues in Human Genetics,* New York, Plenum, 1973.

stigmatization of persons who have genetic diseases. Not only have we come to use the term "carrier" for ones whose offspring may be at risk for some genetic disease, thereby invoking folklore figures like Typhoid Mary and such traditional cultural terms of opprobrium as "bad seed," but we are also coming to accept the principle that those who are genetically diseased ought not to reproduce; by extension of the "carrier" metaphor, sterilization becomes a form of quarantine. We thus slide into social policies and attitudes that ought at the least to be squarely faced, independently debated, and adopted, if at all, both with the justification of a solid cost–benefit analysis and with full knowledge of the price paid in the currency of individual liberty and self-determination.

The identification of a disease with its genetic component, since it displaces the patient's complaint as the primary determinant of the occurrence of the disease, has other potential political and social consequences that raise grave concerns. One effect of separating disease from the diseased one's complaints is that the finding of disease can be made on the complaints, actual and anticipated, of others. The controversy surrounding the "XYY syndrome" serves as a case in point. An unproven, low statistical association of criminality and the occurrence of an extra Y chromosome in males has been suggested. Walzer and Gerald of Harvard Medical School initiated a long-term study of newborn XYY males, in part to determine the degree of genetic (i.e., chromosomal) contribution to behavior problems. But they have been quoted as holding that the XYY chromosome pattern "is a 'disease,' . . . and . . . children who have it are entitled to medical treatment just as they would be for any other disease"—this despite their admission that while "some XYY children are 'hard to handle,' others are 'perfectly fine'."[29] The associated syndrome seems to have at best only an indirect relation to the complaints of XYY individuals; rather it is the complaints of parents, teachers, and those who have been victimized by XYY criminals that constitute a large part of the syndrome. And, XYY individuals are all classified as diseased, irrespective of whether they are "hard to handle" or "perfectly fine," because they possess the genetic basis for the disease. It appears evident that there is an enormous environmental component in the determinants of the antisocial behavior of those who are hard to handle; how does calling the XYY individual diseased on the basis of his chromosomal pattern enhance our understanding of why the behavioral syndrome occurs? Given that the

---

[29]Barbara Culliton, "Patient's Rights: Harvard is Site of Battle over X and Y Chromosomes," *Science* **186,** 715–717 (1974).

incidence of the correlation is low, it looks as though the identification of the disease with the genetic component only serves to direct attention to a relatively unimportant component of the total cause. And, so directed, we are led to the utterly useless stigmatization of XYY individuals as all incurably diseased.[30]

Finally, let me cite the case of some research that is in progress into the genetic components of cancer, and point to two possible ways of handling the outcome. It has long been known that susceptibility to bronchial carcinoma in cigarette smokers is varied; the occurrence of lung cancer in individuals who have a similar history of rate and duration of smoking is markedly dissimilar (a fact which has bolstered the claims of cigarette manufacturers that it has not been proven that smoking causes cancer). In recent years, the National Institutes of Health have supported research into the basis for such variability. Recent reports idicate that the variability centers about low-, intermediate-, and high-inducible aryl hydrocarbon hydroxylase activities, with susceptibility to bronchogenic carcinoma being associated with the higher levels of activity; these levels of activity are in turn thought "to result from two alleles at a single locus with the three (levels of activity) representing homozygous low and high alleles and the intermediate heterozygote."[31] This and other research has prompted one researcher to observe: "The studies in the journal by Kellerman and his associates suggest that the measurement of BP- (benzopyrene) metabolizing enzymes in human lymphocytes may allow us to detect members of the population who are uniquely sensitive, or who are resistant to carcinogens in cigarette smoke, and further studies along these lines are to be encouraged."[32]

As I view it, the question is not whether such research ought to be continued, but what is to be done with its results. Will it, for example, prompt public concern for those individuals who are carcinogenically sensitive to cigarette smoke, and reinforce legislation to protect them from unintentional exposure to tobacco smoke by restricting the sale and private use of tobacco cigarettes to individuals known to be resistant? Or, will the results of these researches be used to identify bronchial cancer as a manifestation of a genetic disease, therby render-

---

[30]D. Borganokar and S. Shah, "The XYY Chromosome Male—Myth or Syndrome?" *Proc. Med. Genetics* **10,** 135–222 (1974).

[31]G. Kellerman, C. Shaw, and M. Luyten-Kellerman, "Aryl Hydrocarbon Hydroxylase Inducibility and Bronchogenic Carcinoma," *New England J. Med.* **289,** 934–937 (1973).

[32]A. Conney, "Carcinogen Metabolism and Human Cancer," *New England J. Med.* **289,** 971–973 (1973).

ing the individual responsible for avoiding, as best as one can, the noxious substance? Depending on how the knowledge resulting from the research is characterized, what many regard as a matter calling for regulation in the public interest may become a matter involving only prudential considerations for individuals having a genetic disease— despite the fact that bronchial carcinoma is a condition resulting from the interaction of genetic and environmental factors. In light of the valuational components of the criteria listed earlier for the use of "genetic disease," the socialization of the anticipated results of this and similar research should be highly instructive.

To sum up, I have been talking about shifts in the development and application of the notion of disease which have a strongly antihumanistic, anti-individualistic implication. From its original denotation of a dis-ease of some individual, the concept has progressively been sharpened and shaped to include, first, causally related congeries of symptoms and signs, and then, inclusion of specific causal mechanisms of their production as individuators of diseases. I believe that each of the shifts involves a selection among alternatives, and I have argued that the selection among the set of possible causal components is a choice that reflects a decision (perhaps unwitting) to locate responsibility for disease within the individual rather than in environment. This can be seen as directing not only research and treatment toward the individual, but also as directing our conception of causal responsibility away from environmental factors that can be controlled only through social resolve, and locating it within individuals. As it becomes so located, the individual becomes laden with the responsibility to avoid the manifestation and spread of the disease.

The disbenefits for the typology "genetic disease" seem to me to far outweigh any benefits that accrue to research and therapy from the use of the term. If, nonetheless, it is viewed as desirable to have some indication of a disease's etiology encapsulated, if not in its identification, at least in some systematic form of reference, it may be valuable to discriminate between different types of the same disease (specified in terms of an associated syndrome or family of syndromes) in terms of differential causal analyses. But, lest we repeat there the sins outlined above and once again fall into the grip of the simplistic equation of "one disease, one unit cause," we would do well to develop some sort of characterization of the whole cause of a disease in terms of a density function that would provide some specification of the relative contributions made by genetic, and by various types of environmental, factors. In that way, we could perhaps preserve the virtues of the present system of classification while avoiding the nomenclature of *"genetic* disease."

Protecting the Unconceived

Robert Baker

Union College, Schenectady, New York

In recent years a number of public health officials have proposed statutes requiring mandatory genetic screening—and, in some cases, eugenic sterilization—to various governmental and legislative bodies. They justify these proposals by citing the need to protect the interests of society and of the as-yet-unconceived person who, if born, would suffer from a serious genetic defect. Perhaps the most articulate spokesman for the latter position is Professor Michael Bayles, who holds that

> Given the possibility of genetic control, society can no longer risk having genetically disadvantaged children by leaving the decision of whether to have children to the unregulated judgment of individual couples. Some social regulations with respect to genetic screening and, perhaps, eugenic sterilization are needed. While potential parents have interests of privacy and freedom in reproductive decisions, the social interests in preventing the suffering and inequality of possibily defective children may outweigh them in certain types of cases.[1]

In another essay[2] Bayles develops this position further, arguing that in certain kinds of cases—cases in which if an unconceived human were to be conceived and born its existence would be worse than nonexistence—there is a good reason for laws preventing conception. A simplified version of the argument that Bayles offers is the following:

---

[1]Michael Bayles, "Marriage, Love, and Procreation," in R. Baker and F. Elliston, eds., *Philosophy and Sex*, Buffalo, Prometheus Books, 1975, p. 199.

[2]Michael Bayles, "Harm to the Unconceived", *Phil. Public Affairs* **5**, 292–304 (1976). The argument under consideration is weaker than the one Bayles actually develops in this essay, but he and I agree that the stronger argument is sound only if the weaker is.

BI   There is a good reason for laws that prevent people from harming other individual persons.

BII   If A conceives and brings into existence someone, B, whose life is worse than nonexistence, A harms B.

BIII Therefore, there is a good reason for laws preventing A from conceiving and bringing B into existence.

In this essay I shall consider whether the argument is sound and whether, even if it is sound, it provides us with a sufficient reason for enacting compulsory eugenic-sterilization statutes.

How can an argument provide us with a good reason for legislating nonconception without providing a sufficient reason? Clearly, because other good reasons override the consideration in question. Thus, since Bayles' argument is formulated in terms of harms, there would be insufficient reason for the legislation in question if its implementation would generate more harms, either to the conceived or the unconceived, than the behavior in question would generate were it left uncontrolled by legislation. This is, in fact, one of Jeremy Bentham's principles of legislation—that a proposed statute is morally "unprofitable" if "the mischief it would produce would be greater than what it would prevent."[3] If, therefore, we could show that programs of mandatory eugenic screening and compulsory sterilization are "unprofitable" in Bentham's sense of the term, we would have established that Bayles' argument does not provide a sufficient reason for a mandatory eugenic screening–sterilization program.

Basically there are four ways in which Baylesean legislation is unprofitable: through misapplication, through misdiagnosis, by engendering resistance, and by violating the interests of the healthy unconceived.

The harms of *misapplication* and *misdiagnosis* might be said to be empirically-tested, historically-proven methods for engendering the moral failure of eugenics statutes. *Misapplication* arises, in the view of some speaker S, if a compulsory eugenics statute defines some condition C as a genetic defect requiring mandatory sterilization when S does not believe it to be a genetic disease or defect. The Nazis, for example, defined being a schizophrenic, a Romany, or a Jew as conditions requiring eugenic sterilization and/or eugenic "euthanasia." Since few Americans find these definitions acceptable, the Nazi program is generally regarded as a gross misapplication of the ideals of eugenics.

---

[3]Jeremy Bentham, *Introduction to the Principles of Morals and Legislation*, Oxford, 1789, Chapter XIII, sec. III, case 3; reprinted in Mary Peter Mack, ed., *A Bentham Reader*, New York: Pegasus Books, 1969, p. 120.

Misdiagnosis arises when both S and a eugenics statute are in agreement that some condition, C, is in fact a disease or defect, but make a mistake either in identifying which individuals have the disease, or in regarding the disease as heritable. Imbecility, for example, is generally regarded as a defect. But the most famous imbecile in American jurisprudence, Carrie Buck, plaintiff in *Buck vs. Bell*,[4] was not, in point of fact, an imbecile. For, although Carrie was found to have a mental age of only nine years on the Binet test, and so was regarded as mentally defective according to the criteria for normal intelligence then employed, these criteria were later proved to be inadequate, and on the revised Stanford-Binet scale, Carrie's intelligence would have been normal for a white southern female of the time. The revisions in the test norms, however, were too late to save tens of thousands from suffering Carrie's fate—involuntary sterilization.[5]

Unlike misdiagnosis and misapplication, history has not yet established *resistance* as a proven reason for regarding eugenics legislation as morally unprofitable. It is, nonetheless, clear that widespread resistance to an otherwise sensible and beneficial policy can render the policy worthless. Even those who once believed so fervently in the virtues of prohibiting the sale of alcoholic beverages later came to recognize that prohibition, under conditions of resistance, could be counter-productive. That sterilization is the type of issue which can create such resistance is beyond dispute (one need only consider the role that resistance to forced sterilization played in the recent defeat of Indira Ghandi's government). What is not, perhaps, as immediately evident is how small a measure of resistance would be required to render a sterilization policy counterproductive. Consider, for example, the one case of a genetic defect that Bayles mentions as a possible candidate for eugenic control—Tay-Sachs disease. The disease (like most candidates for eugenic control) is exceptionally rare. It is most common among American Jews of Ashkenazic descent, and even in this population the frequency is only 1 per 3600 live births. If, therefore, resistance to a eugenics policy were to lead to an increase in infant mortality and/or retardation in more than 1 in 3600 births, the policy would have to be regarded as "unprofitable." Thus any significant prospect that a disease

---

[4]197 U.S. 11 (1927).

[5]Mark H. Haller, *Eugenics: Hereditarian Attitudes in American Thought,* New Brunswick, Rutgers University Press, 1963. For other historical material on eugenics, see C. P. Blacker, *Eugenics in Prospect and Retrospect,* London, Hamish Hamilton Medical, 1945; K. Ludmerer, *Eugenics and American Society,* Baltimore, The Johns Hopkins University Press, 1972; and M. Woodside, *Sterilization in North Carolina,* Chapel Hill, University of North Carolina Press, 1950.

will produce resistance to eugenics statutes (e.g., a refusal to give birth under medical supervision) should be regarded as likely to be counter-productive.

Unlike the first three scenarios for moral unprofitability, the fourth, *violating the interests of the healthy unconceived,* is almost entirely a priori—although it does rest on some familiar facts of Mendelian genetics. To use the oversimplified language of high-school biology texts, genes are paired and come in two varieties: dominant *(G)* and recessive *(g).* For any trait under the control of a gene pair, a person receives one gene from each parent. Assuming trait G^* is controlled by gene G, and g^* by g, a person will have the trait (or phenotype) G^* (e.g., idiocy) if the inherited gene makeup (genotype) is GG or Gg (since G dominates g); that individual will have the phenotype g^* only when gg is inherited. Thus, if two people mate and one has genotype GG (i.e., if one is *homozygous* for the dominant gene), since all of their offspring will inherit at least one G, all will be phenotypically G^*. If two Gg genotypes (heterozygotes) mate, they should produce three G^* phenotype children for every one g^*; about half of their children should be genotypically heterozygotes *(Gg),* one quarter homozygous for the dominant trait *(GG),* and one quarter homozygous for the recessive trait *(gg)*—only the last group will have the phenotype g^*.

Now suppose that a legislature is wondering whether a trait is sufficiently dysgenic to warrant eugenic sterilization. Since Bayles' criterion for compulsory intervention (in premise BII of his argument) is only applicable to traits that render existence worse than nonexistence, no dominant G^* trait of competent married adults could satisfy the criterion. For a child would inherit such a trait if and only if at least one parent were both genotypically G and phenotypically G^*.[6] And *since this G^* parent is competent, adult, procreative, and presumably nonsuicidal, his very existence provides what might be called existential evidence of a fairly compelling nature that a G^* existence is not worse than nonexistence,* Hence, on Bayles' criterion, dominant G^* defects are not susceptible to eugenic control.

Neither are recessive defects. For our purposes recessive defects can be thought of as those resulting from matings in which one or more of the parents are homozygous *(gg)* for the defect *(g*),* and those resulting from matings in which neither of the parents is a homozygote. The former case provides precisely the same existential evidence that

---

[6] Genetic theory and fact cited throughout this paper have been checked against L. L. Cavalli-Sforza and W. F. Bodmer, *The Genetics Of Human Populations,* San Francisco, W. H. Freeman and Co., 1971 and/or J. A. Fraser Roberts, *An Introduction to Medical Genetics* (sixth edition), London, Oxford University Press, 1973.

proved dominant traits insusceptible to Baylesean regulation. For, once again, the child only inherits the defect, g^*, if at lest one of his parents is living a g^* existence—which presumably would suffice to show that such an existence is not worse than nonexistence.

The case of heterozygote carriers of recessive genetic defects is a bit more complex. Again, it may be analyzed as two subcases: that in which the heterozygote carrier of a g^* defect *(Gg)* is mated with a normal person *(GG)*, and that in which both mates are heterozygotes *(Gg)*. In the former case the healthy G will dominate the defective g; hence none of the children will inherit the defective g^* phenotype, and there will consequently be no grounds for eugenic regulation. In the latter case, according to the normal Mendelian rules, even though neither parent would display the defective phenotype, one quarter of their children would display it, and three quarters of their children would inherit a defect-carrying genotype. The situation is summarized on the following chart.

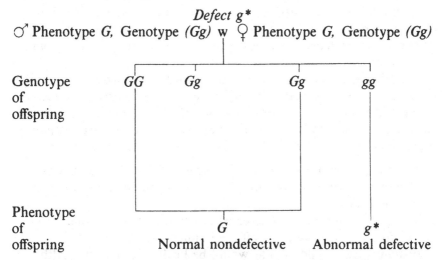

Defect g^*

♂ Phenotype *G,* Genotype *(Gg)* w ♀ Phenotype *G,* Genotype *(Gg)*

Genotype of offspring: GG Gg Gg gg

Phenotype of offspring:

G
Normal nondefective

g^*
Abnormal defective

Clearly, since neither of the parents has the defective phenotype g^*, their existence and procreative activities cannot provide existential evidence that g^* is not a condition worse than nonexistence. Thus we have here the type of case in which it would be prima facie reasonable for a legislative body to protect the unconceived by constraining the procreative activities of their potential parents. It is, in fact, the common genetic pattern for the transmission of genetic defects which, like Tay-Sachs disease, can be reasonably said to constitute an existence that is worse than nonexistence. And unlike the previous cases analyzed, Bayles' criterion is properly applicable to this case.

The application, however, is unprofitable. Why? Because sterilization is indifferent to the quality of conceptions it prevents. It can only
prevent the conception of the phenotypically unhealthy genotypes if it
also prevents the conception of the phenotypically healthy genotypes.
Therefore, whatever benefits sterilization might bring to the defective
unconceived must be weighed against the disadvantages that the
healthy would suffer were they to be deprived of their chance at existence. In the case under consideration, given the Mendelian rules (illustrated in the diagram above), the healthy will outnumber the defective
by a ratio of three to one. Hence sterilization to protect the unconceived
appears to be thrice more harmful than beneficial—in sum, unprofitable
and counterproductive.

Morally, the situation seems parallel to one in which an individual
may bring it about that a newborn Tay-Sachs child should not come
into existence *if* it is also willed that three other healthy potential
newborns should not come into existence. And it seems counterintui-
tive, indeed, unwarranted on any major moral theory (including
utilitarianism), to will the non-coming-into-existence of the three
healthy children in order to secure the non-coming-into-existence of the
one unhealthy child.

Nor does a detailed analysis of specifics improve the situation.
Consider, for example, the theoretical case of 64 married couples in
which both partners are heterozygote *(Gg)* for Tay-Sachs disease
(which, you will recall, is a recessive trait, *g**). In all oversimplified
probability, were each of the couples to have three children:

27 couples would not have any children affected by Tay-Sachs
27 couples would have two unaffected children and one affected
    child apiece
9 couples would have two affected children and one normal child
1 couple would have three affected children
In sum, of the children 144 would be healthy and 48 unhealthy
    (note the 3:1 ratio); of the couples, 54 would have more healthy
    children than unhealthy children, and only then would have
    more unhealthy than healthy children.[7]

Thus sterilization would be grossly unfair to the 27 couples who would
have no defective children at all; would likely be disadvantageous to the

---

[7]These figures are drawn from p. 307 of Robert Murray's excellent article "Ethical
Problems in Genetic Counselling," *Ghanaian Med. J.* **12,** 304–309 (1973). See also
Murray's "The Practitioner's View of the Values Involved in Genetic Screening and
Counselling" in *Ethical, Social, and Legal Dimensions of Screening for Human Genetic
Disease,* **10** (6), 185–199 (1974).

27 couples who would have more healthy than unhealthy children; and might only be beneficial to the fewer than one couple in five who would have more defective children than not. As for the potential children, if the average expected life span of a normal person is three score years, and if the average expected life span of a Tay-Sachs child is three years, then 144 of the healthy unconceived are being deprived of some 8640 person years of an existence better than n in order to save some 48 unconceived amaurotic idiots from 144 years of existence at level less than n. And to sacrifice so much to gain so little is, to use Bentham's language, "unprofitable."

In sum (see chart below) a Baylesean sterilization program is only permissible in cases where the defect is carried by a recessive trait, and then only for mated heterozygotes and, in this the only relevant case, it is unprofitable because of the 3:1 healthy:unhealthy ratio.

I Case:	*Dominant Defect*	*Genotypes*
	One or more partners dominant for defect G^*	GG w GG
	At least one parent has defective G^* phenotype	GG w Gg
	All children will have defective G^* phenotype (except in the last case, where 50% will be normal)	GG w gg
		Gg w gg
	Reason for Exclusion Existential evidence, G^* not worse than nonexistence since one or more parents have defective G^* phenotype	

II Case: *Recessive Defect,* two recessive homozygotes mate. Both parents have defective g^*
All children have defective g^* phenotype
Reason for Exclusion
Existential evidence g^* not worse than nonexistence since both parents have defect g^*

$$gg \quad \text{w} \quad gg$$
$$gg \ gg \ gg \ gg$$

III Case: *Recessive Defect,* only one partner heterozygous for defect g^*
No parents have defective phenotype g^*

$$GG \quad \text{w} \quad Gg$$
$$GG \ Gg \ GG \ GG$$

No children have defective
phenotype $g*$
Reason for Exclusion
None of the children inherit
defective phenotype

IV Case: *Recessive Defect,* both partners Gg w Gg
heterozygous for defect $g*$
No parents have defect $g*$ GG Gg Gg gg
$g*$

*1 Child in four should inherit
defective*
Reason for Exclusion   phenotype $\underline{G}$
Moral unprofitability of eliminating.
three healthy unconceived to save   phenotype $g*$
one unhealthy unconceived from
conception

It is no small measure of the depth of Bayles' analysis that he, uniquely among eugenicists, has proposed a justification of eugenics sterilization that is sensitive to the problem of the healthy unconceived. He puts the question "can one harm a person by failing to conceive and bring about his existence?"[8] and answers that one can *not*. Therefore, since the prevention of the conception of the healthy unconceived does not constitute any harm at all, preventing such conceptions by sterilization could not possibly harm thrice more than it helps. To quote Bayles, "The harm condition for the application of the principle [BII] is not met when one fails to conceive . . . one can not provide a morally neutral analysis, one that does not presuppose a duty to procreate, to show that a person is harmed by not conceiving. The non-existent can not be made worse off without being brought into existence. Perhaps one can deprive them of benefits, (including life), but one can not harm them without bringing them into existence."[9]

How does Bayles go about proving the asymmetry that *bringing* someone into existence can be a harm, but *preventing* someone from coming into existence can not be?

From the fact that one would be better off were one to act in a certain way, it does not follow that failure to so act harms him. Even if it would be good for Jones (promote his interests) for Smith to give him $500, it does not follow that in not giving Jones $500 Smith harms him. To ignore

[8]Bayles, "Harm to the Unconceived", p. 294.
[9]*ibid.*, p. 299.

this is to collapse the distinction between harm and non-benefit. While this distinction is very difficult to articulate precisely, it needs to be retained; otherwise there would be no conceptual room for non-obligatory gifts.[10]

The conceptual underpinnings of Bayles' analysis are simple enough. Sometimes A's actions have an effect on B such that B is better off than he or she would have been had A not acted; such actions benefit B. Sometimes A's actions make B worse off (by, in Bayles' words, "either unreasonably risking adversely affecting, or . . . adversely affecting" B's net interest[11] such actions harm B. Some of A's actions leave B neither better nor worse off; when these actions are alternatives to helping actions they are nonbenefits. Now suppose that A has a choice between three alternative courses of action: C^1, C^2, and C^3. C^1 will make B better off; C^2 will make B worse off; and C^3 will leave B neither better nor worse off. Since A could act to make B better off (C^1), if A chooses C^3 instead, the choice will not benefit B—it will be a nonbenefit. And, as Bayles' example of Smith's decision not to give Jones a present of $500 illustrates, conferring a nonbenefit, such as C^3, is not a harm such as C^2. For harms make a person worse off, but nonbenefits leave the situation unchanged.

Conception, the argument continues, is, like Smith's $500, a gift —a rather superlative gift to be sure, but a superlative gift is still a gift. And gifts, by their very nature, are benefits; hence, since to fail to give a gift is merely to fail to confer a benefit, "to fail to conceive and bring about someone's existence" is merely a nonbenefit and, as such, can not possibly be a harm. Therefore, it follows on Bayles' analysis that the nonconception of the healthy unconceived caused by sterilization does not constitute a harm to them.

This analysis presupposes, of course, that conception is properly regarded as a benefit. Yet, oddly enough, even though his entire paper rests on this assumption, Bayles never gives us any reason to believe that this beneficial notion of conception is correct. Is it? To answer this question we must have some way of identifying an unconceived individual (or a class of such individuals) and of determining what actions are adverse for, or advantageous to, them. Suppose we think of an unconceived individual U as one possible karyotype (i.e., a set of 46 chromosomes arranged as 23 pairs—*1, A,B; 2, C,D . . . 23, XY*—which constitute a given human individual's genotype). Now there seems to be a significant difference between a possible human being and an uncon-

[10]*ibid.*, p. 298.
[11]*ibid.*, p. 293.

ceived human being; a possible human, *P,* is simply any karyotype that is combinatorially possible given the total set of human genes. Whereas an unconceived human, *U,* is someone that a particular mating couple, *M,* might conceive in a fertile act of intercourse. *P* and *U* differ because the unconceived are designated by reference to a couple *M,* whereas possible persons—a class that includes the unconceived—are not so limited. In general, Bayles, indeed most eugenicists, are talking about the unconceived, since they are always dealing with the possible results of matings of particular couples—those who carry defective genes.

Having made some rough determinations about how to locate *U,* we can now ask whether *U*'s conception should be regarded as a benefit to *U* or, contrapositively, whether *U*'s nonconception would be a non-benefit to *U*? It is exceptionally difficult to find a reasonable way of answering this question. Presumably, if *U* anticipated an existence better than nonexistence, the conception and birth would be a good thing. But that would not make it a benefit, since to be a benefit in the sense in which Bayles uses the term is to partake of a relation involving someone *(A)* whose action makes *U* better off than *U* would otherwise have been.

Perhaps we would have better luck if we tried to determine whether a *U* anticipating an existence better than nonexistence is *harmed* by an act of sterilization. For that, after all, is the real point at issue—*whether sterilization by preventing the conception of three of the healthy unconceived constitutes a harm that outweighs the benefit of preventing the birth of the one unhealthy unconceived.* I find it hard to escape the conclusion that for a healthy *U* any action that lowers the probability of existence satisfies Bayles' own characterization of a harm —i.e., it has an adverse effect on *U*'s net interest. Clearly *M*'s steriliza-tion, indeed any act of contraception, lowers the probability of *U*'s conception. Hence it would seem to follow that if the unconceived can be said to have interests and so to be harmed, acts of sterilization that prevent the conception of the healthy unconceived are harms to these unconceived. Given this model of the interests of the unconceived, Bayles' analysis is wrong.

Is there a reasonable analysis which permits the conclusions Bayles is trying to draw—i.e., an analysis that will permit him to argue that eugenic sterilization will help the unhealthy unconceived, but not harm the more numerous healthy unconceived? It is difficult to imagine how such an analysis might be formulated, but it is perhaps even more significant to note that even if Bayles' analysis of healthy conception as a benefit were sound, eugenic sterilization would still constitute a harm to the healthy unconceived. Why? Because even if healthy conception should be regarded as a benefit that parents, M, bestow on an unconceived, U, to act to prevent U from receiving that gift, once M has undertaken to give it, constitutes a harm to U. It is one thing for M, or Bayles' Mr. Smith, to decide not to give a gift, and quite another for some third party, the State, to prevent Jones or U from receiving the gift. The former case involves a nonbenefit that harms no one; the latter involves a theft that harms everyone—for stealing the present M has decided to give U harms both M and U (i.e., both are made worse off than they would otherwise have been). Hence, whether or not Bayles is correct in believing that conception is properly regarded as a benefit, eugenic sterilization compelled by a third party, the State, must be regarded as a harm to the healthy unconceived.

The very idea of "protecting the unconceived," the idea that underlies some first-wave and most second-wave eugenics proposals, presupposes that one can form some reasonably clear idea of the rights and/or interests of the unconceived. And we have, throughout this essay, not only written as if it were reasonable to do this, but even proposed formulas for determining the rational expectations of the unconceived. Yet it is extraordinarily difficult to identify an unconceived U. Consider, we have identified U as one of the specific karyotypes genetically possible from the mating of a given couple M. But when does U come into existence? When the partners in M come into existence? When they meet? When they mate? When they contemplate coition? During their first act of coition? During the act of coition in which U comes to be conceived? Or only (retroactively) after U's conception? The answers to these questions make a significant difference. For it is only if one thinks of U as preexisting some copulatory acts of the couple M that sterilization can be held to be a benefit or a harm to U. And who is U? We have regarded U as a karyotype, but although a karyotype is an individual genotype, individual genotypes are not human individuals. For identical twins, although they are two distinct human individuals, are one and the same individual genotype.

Given the metaphysical complexity underlying the determination of the interests of the unconceived, it is small wonder that Lord Kilbrandon,[12] Tedeschi,[13] and courts in New Jersey[14] and New York[15] have

---

[12]See Lord Kilbrandon, "The Comparative Law of Genetic Counselling" in Bruce Hilton, *et al.*, eds., *Ethical Issues in Human Genetics*, New York, Plenum Press, 1973.

[13]See G. Tedeschi, "On Tort Liability for 'Wrongful Life'," *Israel Law Rev.* 1, 513–538 (1966).

[14]*Gleitman vs. Cosgrove*, 49 N.J. 22, 227 A.2d 689 (1967).

[15]*Stewart vs Long Island College Hospital*, 35 A.D.2d 531, 313 N.Y.S.2d 502 (1970).

concluded that "it is impossible to make such a determination."[16] I should prefer to argue that such determinations are not so much impossible as so arbitrary as to be without moral or legal force. Be that as it may, it does seem reasonable to argue, contra Bayles and others who justify eugenics programs by appealing to our duty to protect the unconceived, that the unconceived have no moral or legal rights.

One might, in the light of the analysis undertaken above, put the argument against compulsory eugenic sterilization undertaken in order to protect the unconceived as follows. Either the unconceived have determinable interests or they do not. If they do not, then there are no grounds to override the known rights and interests of the conceived to protect them, and hence a eugenics policy to protect the unconceived is unjustifiable. If the unconceived do have rights and/or interests, then the healthy as well as the unhealthy unconceived must be allowed to have such rights and/or interests. And if they do, then, since there will always be more healthy than unhealthy unconceived, no eugenic sterilization policy will be justifiable. Thus, whether or not the unconceived have rights and/or interests, no compulsory eugenic sterilization policy can be justified on the grounds of protecting the unconceived.

---

[16]*Gleitman vs Cosgrove,* 227 A.2d 689, at 692 (1967).

Comments on "Protecting the Unconceived": Butchers, Bakers, & Candlestick Makers

Michael D. Bayles

Department of Philosophy, University of Kentucky, Lexington, Kentucky

A deplorable practice of obscure philosophers is to claim their critics misunderstand them. In these comments I confess to being deplorable and obscure by claiming I have been misunderstood. Fortunately, the view Professor Baker attributes to me is, with some qualifications, more defensible than that which I in fact hold. Since it is a special case of the view I accept, if it is wrong so is my view.

In "Harm to the Unconceived," I criticized the adequacy of John Stuart Mill's harm principle for justifying eugenic sterilization and other legislation to protect the unconceived.[1] Mill claimed that the only good reason for laws limiting liberty is to prevent people harming others. Two elements are necessary for the (private) harm principle to support legislation: (1) the prevented conduct must affect another particular person, and (2) it must harm that person. Mill denied that popular morality or a person's own well-being provided reasons for laws limiting a person's liberty.

The harm principle, I claimed, rarely supports legislation to prevent people from having genetically defective children. If a defective infant is conceived and brought into existence, then there is an individual person, but unless that life is worse than nonexistence, the person is not harmed. Moreover, one cannot say a defective infant whose life is better than nonexistence has been harmed by being born instead of

---

[1]Michael D. Bayles, "Harm to the Unconceived", *Phil. Public Affairs* **5** 292–304 (1976).

a normal child, because a normal child would not be him but a different person. If conception of a child is prevented, there is no particular person; and even if there were, it would not be harmed because it would be no worse off than before—not existing in either case.

Having thus shown the harm principle to be of very limited application, I suggested a different principle. Its premise is that if one chooses to produce a good for others, it is wrong to produce one below some acceptable minimum. The principle is as follows: There is a good reason for legislation to prevent the birth of persons who would lack substantial capacity to achieve or take advantage of a quality of life of level n or whose existence would decrease the number of people who might live with a quality of life at that level. Level n is the acceptable minimum. (The second half of this principle was designed to set a maximum size to population and is not relevant to the present topic.)

By taking my principle to be a version of the harm principle, Professor Baker has managed to interpret my article as a defense of the use of Mill's principle to support eugenic legislation. Actually, my principle was and is intended to be much broader than the harm principle, although it includes the harm principle as a special case. If the quality of life of level n is such that persons whose lives are not worth living fall below it, then my principle provides a reason for legislation to prevent the birth of such persons. However, my principle requires only that classes of persons, not particular ones, be harmed. Consequently, I am willing to defend the argument Baker attributes to me, as its major premise, BI, is a special case of my principle.

BI   There is a good reason for laws that prevent people harming other individual persons.

BII   If A conceives and brings into existence someone, B, whose life is worse than nonexistence, A harms B.

BIII   Therefore, there is a good reason for laws preventing A from conceiving and bringing B into existence.

For a defective infant to be harmed, its life must be worse than nonexistence. Preventing the birth of unconceived people does not benefit them; it does not make them better off. At most it prevents them being harmed, being worse off because their lives would be worse than nonexistence. One values or prefers nonexistence to a miserable existence; the criteria for making such judgments are another paper and not crucial here.

Baker's criticism of the harm argument for eugenic sterilization laws is that the laws would produce more harm than they would prevent. If he is correct, then such laws should not be adopted, and the harm principle, properly interpreted, does not support them. Properly interpreted, the harm principle supports only legislation which, *everything considered*, prevents harm. Baker suggests four ways in which eugenic sterilization laws may produce harm. *(1)* They may be misapplied. *(2)* Genetic conditions may be misdiagnosed. Whether misapplication and misdiagnosis are likely depends upon the particular law proposed and the context of its use. Although these problems have been frequent in the past, the ability correctly to diagnose transmissible genetic disease has greatly increased during the last 20 years, and stringent due process safeguards may prevent most misapplication.

(3) There may be resistance to the sterilization. Baker suggests that very little resistance is needed for the harm produced by it to outweigh that prevented; for Tay-Sachs, if women give birth without medical supervision, a frequency of more than 1 defect per 3600 live births would produce more harm than sterilization would prevent. However it is not just the rate of defects that must be considered, but also the frequency of such births. Even if the rate of defects in unattended births were 100% higher, if less than one-third of the target population were to engage in resistance there would be a net decrease in children born with defects. Moreover, most people would not have reason to resist, because they would not be at risk of having defective children and thus sterilization. Of those who are, the vast majority do not wish to risk defective children. The Indian resistance to mandatory sterilization is not pertinent, because the proposal there was to sterilize people regardless of the risk of defective children. Thus, although resistance may theoretically provide a reason against compulsory sterilization, its importance cannot be determined a priori, and I doubt it would be large enough to block application of the harm principle.

(4) Baker's main claim is that eugenic sterilization always produces more harm than it prevents, because it harms the healthy unconceived. For a number of reasons, his argument on this point is not convincing. (a) Baker seriously limits the scope of his argument by the category of persons he takes as potential candidates for eugenic sterilization. He considers sterilization only of competent, married adults. Thus, his argument is inapplicable to people who have genetic defects sufficiently serious to render them incompetent and leaves completely untouched use of the harm principle to support judicial decisions such as that on the presumed facts in *Buck vs Bell*.[2]

(b) Baker argues that if a competent, married adult manifests a defect, then inheriting the defect cannot be harmful. This argument is

---

[2]274 U.S. 200 (1927).

unsound. Even though a person manifesting a defect is competent, the defect may be serious enough to render the lives of *most* persons with it worse than nonexistence. Genotypes are variously expressed, as is evidenced by sickle cell disease. Suppose that 1 in 100,000 persons with a disease or defect is competent, but the rest have miserable lives. The fact that a particular person is competent does not show that its off-spring with the defect would not very probably have a life worse than nonexistence.

(c) Baker uses what has been called the milk production model of utilitarian calculation—the more milk or life of higher quality, the better. Thus, he calculates how many lives or life-years of good or bad quality would exist were sterilization not performed. He concludes that since there is more good than bad life, one ought not sterilize. This argument rests upon the criterion of total utility, which has been shown unacceptable by a number of authors.[3] I clearly rejected it in my original paper, writing, "There is no obligation to produce as much good as possible."[4]

The fundamental issue is whether one can or should take into account the interests of the unconceived who are not born. After presenting the above argument against my view, Baker proceeds to doubt whether it makes sense to consider the interests of the unconceived, both because they do not have interests and because particular unconceived persons cannot be identified. It is disheartening to have one's arguments presented as a criticism of oneself. One appears to be impaled on the Cretan liar paradox; if one is correct, then one is wrong. First, I wrote, "Non-existent persons do not have any interests which may be adversely affected."[5] It is not that their interests may only be favorably affected; they simply do not have any interests at all. Second, I denied that one can identify particular unconceived persons. I wrote that R.M. Hare's proposed description of " 'the person who will be born if these two people start their coitus in precisely five minutes' does not designate anyone in particular."[6] Any description of unconceived persons is like Hare's of the next person to occupy a library carrel, which does not identify a particular person, because any of several

---

[3]John Rawls, *A Theory of Justice,* Cambridge, Mass., Harvard University Press, 1971, pp. 162–163; Peter Singer, "A Utilitarian Population Principle", in *Ethics and Population,* Michael D. Bayles, ed., Cambridge, Mass., Schenkman Publishing Co., 1976, pp. 81–84.

[4]Bayles, *op. cit.,* p. 303.

[5]*ibid.,* p. 294.

[6]*ibid.,* pp. 299, 300. See R. M. Hare, "Abortion and the Golden Rule", *Phil. Public Affairs* **4** 220 (1975).

particular persons might be the next to occupy the carrel. There is a significant difference between definite descriptions referring to the future and those referring to the present or past. Definite descriptions whose defining characteristics have future reference guarantee that one and only one person will satisfy them, but they do not enable one to now pick out who that person may be. Hence, I partially summarized the reasons why the harm principle does not apply to the unconceived as "the condition of an individual person is in fact met only after conception brings about a person's existence."[7]

Analogies may clarify my view about the status of the unconceived with respect to identity, harm, and benefit. Suppose John tells Fred that tomorrow there will be a bottle of scotch in the locked liquor cabinet, because tonight Susan is going to put one in it. At present Fred does not know which particular bottle of scotch will be in the liquor cabinet. One can then imagine a murder case in which an attorney examines Fred to determine whether he can identify this bottle of J&B as the one that was in the liquor cabinet the next day. Our ability to identify the unconceived is no better than Fred's ability to identify the particular bottle of scotch. We may know that a human being will exist, we may even be able to describe it as this woman's next child, but we do not know which particular person it will be. We do not even know which sex it will have.

Baker criticizes me for claiming that conceiving and bringing about a person's existence is to benefit that person. Of course, if that claim is false, Baker's argument about the healthy unconceived collapses, for there is no benefit of which they might be deprived. Although I am not sure I did claim conception is a benefit, one can make sense of it. A slight modification in the liquor cabinet analogy may illustrate the point.[8] Suppose that on the following day John gives the key to it to some unknown person, for example, by taping it with instructions to a door various people pass so that the next passerby can find it. Does John benefit the person who finds the key? The key is necessary to getting the scotch. So if the scotch is good, then John's "giving" the key to whomever finds it benefits that person, although the key is not a value in addition to that of the scotch. Analogously, in the liquor cabinet of life, receiving the key by being conceived and brought into existence is a benefit if the life is worthwhile or valuable, even though being con-

---

[7]*ibid.*, p. 300.

[8]Here the scotch is analogized to the life, whereas in the previous paragraph it was analogized to the unconceived person.

ceived and brought into existence is not a value in addition to the value in one's life.

Moreover, one can distinguish when the key to life is a benefit from when it is a harm. If the scotch is good, then getting the key to the liquor cabinet is a benefit. Failure to get the key would be nonbenefit, that is, failure to receive a benefit. Yet, suppose the scotch is bad, worse than no scotch at all. Then in "giving" the key, John harms (perhaps unknowingly) whoever finds it. Similarly, conceiving and bringing someone into a miserable existence, one worse than nonexistence, is to harm that person.

Finally, one must remember that in these cases of getting good or bad scotch, there is no problem identifying who is harmed or benefited; it is whoever finds the key. When conception and birth occur, there is no problem in identifying the particular person who has been harmed or benefited. If the liquor cabinet key is not given to someone, or if no one is conceived and brought into existence, then there is neither harm nor benefit; there is no one to be harmed or benefited. Thus, if conception is prevented, as by eugenic sterilization, no person exists to whom harm was prevented. But the fact that one cannot identify a particular person who would have been harmed but for the sterilization does not prevent use of the harm principle. The harm principle requires only that when harm occurs, one can identify a particular person who is harmed. For example, in passing a law against murder, the legislature cannot specify the particular persons whose lives will be saved. The harm principle nonetheless supports such a law, because when murder occurs one can identify the particular person who is harmed. Similarly, if a child is conceived and brought into a miserable existence, one can identify the person who is harmed.[9]

Baker contends that the unconceived might be harmed because once potential parents undertake to conceive and bring them into existence, preventing their giving the benefit is harm to the unconceived. However, the reason there is normally harm in such cases is the frustration of legitimate expectations. This consideration does not apply to the unconceived. There are no identifiable individuals involved. Even if there were, they could not have legitimate expectations because they do not and cannot have any expectations at all.

---

[9]Of course, one can also identify a child who is conceived and brought into an existence better than nonexistence. But the harm principle does not provide a reason for producing good, only avoiding harm. Hence, another premise, such as that of total utilitarianism is needed before one has any reason to balance foregone benefits against the harms the way Baker does.

If sterilization does not harm the healthy unconceived, what harm might there then be in eugenic sterilization that would provide a reason against it? Baker's reasons of misapplication and misdiagnosis are not harms to unconceived infants. Why then are they harms to be balanced against those prevented? Baker never tells us, but surely common sense tells us they are harms to the persons sterilized, not to children they might have had.

So the issue in eugenic sterilization is whether the harm to children who would be conceived and brought into a miserable existence would be greater than that to those who would be sterilized. I cannot here argue this point in detail. Instead, I shall present a semihypothetical case in which I think the harm principle would justify involuntary eugenic sterilization. Children affected with the recessive Lesch-Nyhan disease have a compulsion to self-mutilation; they eat their fingers and hands and otherwise mutilate themselves. (They also attack other people.) To prevent such mutilation they must be physically restrained from almost all physical movement. For example, I have been told of an eleven year old boy who was essentially chained to the wall for almost all his life. Moreover, frustrating the compulsion to self-mutilation seems to produce anguish and suffering.

Although it is now possible to detect the disease *in utero* and abort affected fetuses, suppose that were not possible, but one could detect carrier status. Would the restriction on freedom, bodily invasion, and ability to beget involved in involuntary sterilization of one member of a couple who were at risk of conceiving such a child be a greater harm than the miserable life of such a child? I believe the child's life would be worse than nonexistence and that the harm to the potential parents, given that they could adopt children, is significantly less than that which would be prevented. How many, if any, diseases and cases of this sort there might be, I do not know. I only claim that if there are such cases, the private harm principle justifies involuntary eugenic sterilization.

To summarize: the unconceived have no interests and cannot be harmed or benefited while unconceived. Moreover, particular unconceived persons cannot be identified. However, if someone is conceived and brought into existence and has a life worse than nonexistence, that individual has been harmed. A eugenic sterilization law that prevented this from occurring would prevent harm to particular persons, even though to the extent it were successful, like laws against murder, one could not say which particular persons had been protected. Moreover, by preventing the births of persons whose lives would be better than nonexistence, no particular unconceived person is harmed. There is no such person, and in any case it is failure to benefit rather than harm. Thus, eugenic sterilization involves no harm to particular unconceived persons and may prevent harm to particular persons. Any harm in eugenic sterilization is to those who are sterilized. Yet, there are at least conceivable cases in which the harm prevented may be greater than that inflicted and involuntary eugenic sterilization justified by the (private) harm principle.

8

Sterilization, Privacy, and the Value of Reproduction

Joseph Ellin

Department of Philosophy, Western Michigan University, Kalamazoo, Michigan

> *"But the man had hereditary tendencies of the most diabolical kind. A criminal strain ran in his blood, which, instead of being modified, was increased and rendered infinitely more dangerous by his extraordinary mental powers."*
> —Sherlock Holmes on Professor Moriarty, in "The Memoirs of Sherlock Holmes," by Arthur Conan Doyle (1893)

I. INTRODUCTION

In this paper I will discuss some issues associated with the ethics of sterilization. The main issues I shall discuss are whether the capacity to procreate, which of course is destroyed by sterilization, is an important capacity worth preserving, and whether involuntary sterilization violates the right to privacy. Of course, there is some sense in which the capacity to reproduce is worth preserving, otherwise the human race would soon cease to exist; hence the manner in which I shall raise the question is this. Let us assume a person has the option of adopting children as an alternative to producing his or her biologically own children. Then, under these circumstances, is there anything important about preserving the capacity to reproduce? In other words, my question focuses on the matter of preference: is it important that people who prefer to have their biologically own children, rather than adopt, be allowed to do so? I shall argue that it is not.

With respect to the matter of privacy, here we must distinguish between the capacity to reproduce and the procedure of sterilization, which destroys that capacity. I argue that the involuntary loss of the capacity in itself does not violate the right to privacy. Here I differ with the apparent trend of U.S. court decisions. Sterilization, however, is another matter, and I do give reasons why sterilization—at least that procured by present technology, which requires an invasion of the body —might violate the right to privacy.

I should point out that my arguments will not speak to the effectiveness, usefulness, advisability, and so on of sterilization as a technique. There may be cases in which sterilization would be useful in the fight against genetic defects, or in population control, or in other matters; or there might not be. My concern here is exclusively with some of the ethical aspects of the problem.

II. THE IMPORTANCE OF PROCREATION

There is no doubt that many, perhaps most, of us prefer to have our biologically own children rather than to adopt. In asking whether it is important to be able to retain the capacity to create children, I wish to ask first whether this preference is rational. In saying that it is rational to prefer A to B, I mean that there is some good reason why A is better than B. But not every preference that is rational is important. I will therefore say that a preference for A over B is important if the preference is rational and if there is some reason why A is substantially more satisfactory than B. This phrase is meant to suggest that there is some large human interest that A serves and B does not, such that were B to replace A, one's life would be impoverished in a significant way. The point is that not everything which we recognize as valuable is to be thought of as important: only those things whose absence would be seen as impoverishing are important.

A preference cannot be important unless it is rational. It is true that people sometimes say that something is important *to them,* even when they do not have reasons for saying that it is better than the alternatives. For instance, someone might say: it is important to me to know that my children are really mine, even though I do not really have any good reasons for preferring my biological children. But if two alternatives are equally good (i.e., there is no reason why one is better than the other), the choice between them cannot be important, even though the person making the choice may think that it is. Nothing can be important unless there is a reason why it is important, and no alternative can be important when compared to another unless there is a reason why it is better than the other.

Another objection to my criterion of importance is that someone might hold that whether one thing is better than another depends on what the person takes to be important, so that the criterion is circular. For example, if it is important to a person that he or she know that his or her children are really his or her own, then it will be important to that person to be able to have his or her biologically own children, and the preference for this will be rational. But this is only true provided that there is a reason why the knowledge is important. Something does not become important just because someone believes it is important. That reason cannot be simply the fact that the thing is desired or preferred. Preferences cannot be rational just by being preferences: the fact that you like vanilla ice cream better than chocolate may be good enough reason to choose vanilla over chocolate—where you have a choice, it would not be rational to choose the thing you like less—but is not itself a reason to prefer it.

This criterion of rational preference is more adequate than another, more common, criterion. According to this other criterion, a desire is rational if a person would want to act on it, after considering it dispassionately, provided that person were fully informed and free of neurotic compulsions and other psychological afflictions. For our purposes, this criterion is too weak, for it is very likely the case that many of the people who want to have their biologically own children, and who do not want to adopt if they can help it, meet the conditions in the criterion. They simply do not like the idea of adopting someone else's child (as they might see it), and much prefer to create their own. Yet if there are no reasons supporting this preference, we are entitled to call it irrational. This is not to say, of course, that people's preferences should not be respected even if they may be irrational, but where preferences conflict with important goals of public policy, then irrational preferences must be given less weight than rational ones; otherwise society would be thwarted by foolish and even nonsensical desires people might happen to have.

Of course reasons why some things are better than others may vary from person to person: we must not suppose that all reasonable people would have the same preferences in similar situations. To a certain extent, what makes one thing better than another depends on a person's beliefs, theories, hopes, goals, values, and so on. Thus the fact that candidates for public office propose more government spending may be a reason to prefer them if you are unemployed or a socialist, but a reason to prefer their opponents if you are in a high tax bracket or believe in free enterprise. Also, we must not imagine (not that there is much danger of it) that the same things are equally satisfactory to everyone. But neither should we confuse satisfactory with satisfaction. A person might find two things equally satisfactory, yet obtain more satisfaction from one than the other. This is perhaps because 'satisfactory' has to do with the purposes of things, though 'satisfaction' has to do with the pleasure we derive from them. For example, jogging and walking might be equally satisfactory as forms of exercise, yet we might find more pleasure in jogging. But if two things are equally satisfactory, the choice between them is not important: in the example, it is not important whether you jog or walk, although you will enjoy jogging more. (Hence given a choice, it is rational to jog.)

There is no doubt that the reasons people typically give for wanting to have their biologically own children do not have much to recommend them. People want their children to come from "good stock" (theirs); or they want them to have certain personality traits believed to be typical of their own family; or they are afraid of getting a pig in a poke if they adopt children from a family unknown to them. Most of this is based on unproved assumptions at best, and genetic nonsense and racial prejudice at worst; Sherlock Holmes' belief in "criminal strains in the blood" typifies this way of thinking. The argument that people want to know what they are getting has very little validity in today's mobile society in which people are barely aware of the identity, not to mention the characteristics, of their own ancestors, much less their mate's (unlike close-knit societies of earlier times). To the extent to which the partners in a mating are not the products of controlled breeding, supported by adequate records, the good stock argument is invalid.

But is the desire to have one's biologically own children irrational? Some philosophers hold that it is. Michael Bayles for example says: " ... freedom to procreate has little weight [since] it is ... grounded only on a person's unreasoned desires."[1] I agree that if the preference is irrational, then the freedom grounded on it is not important. But Bayles seems to think that unless he can show that the preference is irrational, he cannot show that the freedom is unimportant. Since I have argued that a preference might be unimportant even if it is perfectly rational, I do not have to go as far as Bayles thinks he must go to argue against the unrestrained freedom to procreate. And in fact, Bayles overlooks

---

[1]Michael D. Bayles, "Genetic Equality and Freedom of Reproduction", *J. Value Inquiry* **11**(3), 186–207 (1977),

important facts that might lead one to hold that there are good reasons to retain this ability. Bayles makes the point, correct as far as it goes, that the desire to have children as a symbol of love can be satisfied just as well with adopted children, since love is manifested in the raising and living with a child, not in creating it. He points out that the behavior of one's children is a more important means of identifying with them than their genetic composition and resultant physical characteristics, and children adopted in infancy are no less likely to have valued behavior patterns.

Nevertheless, these arguments overlook certain facts that place the desire to have biological children in a better light. For example, for many people the period during which the fetus is carried by the mother may be a time in which the foundation of the parents' love for the child is created in both parents. Further, some women believe that the experience of pregnancy and childbirth, despite its unpleasant and painful aspects, is a humanly valuable experience they would not willingly forego. Fathers, too, can participate in this experience. Again, a child who is biologically your own is more likely to resemble you physically than one selected at random from the general population, and although this may seem trivial with respect to such characteristics as the shape of one's nose and the cut of one's jaw, there are families in which the family features are a source of pride and family identification in which presumably an adopted child would not be able to share. Furthermore, it is evidently an open question whether there is a hereditary component in intelligence, so that it would not be irrational for highly intelligent parents who want intelligent children to prefer procreation to adoption, since it is possible that their biologically own children might be more likely to be intelligent than a child selected from the general population.

Finally, there is what may be called the spiritual factor. As Leon Kass has said:

"Human procreation . . . is . . . a human activity of the rational will [which] engages us bodily and spiritually as well as rationally. Is there perhaps some wisdom in the mystery of nature which joins the pleasure of sex, the communication of love and the desire for children in the very activity by which we continue the chain of human existence?"[2]

We need not subscribe to the wonderstruck metaphysics of "the wisdom of the mystery of nature" found in this quotation; nor, con-

---

[2]Leon R. Kass, "The New Biology: What Price Relieving Man's Estate?", *Science* **174** 784 (1971).

trary to this author's evident belief, is there any such thing as the activity of procreation: what there is is sexual intercourse, which may be entered into with the hope and expectation that it will result in procreation. Nevertheless the idea of "participation in creation" does seem to have certain metaphysical and spiritual overtones that go beyond mere biological reproduction. The hope and expectation, as well as the subsequent discovery that they have been fulfilled, may for some people have special spiritual value, and hence a preference based on this value would not be irrational. There is in addition a certain kind of pride—which for some amounts to a sense of fulfillment of their masculinity or femininity —about having created a child. Since impregnating or being impregnated are more akin to biological functions than to human accomplishments—more like digestion than like hitting a home run—it may be questioned whether this pride is not misplaced; but where a preference for something is founded in a value such as pride in creation, we cannot, by our criterion, say the preference is irrational.

It will be seen then that, at least to some people, the preference for their biologically own children is a rational preference, since, given their values and beliefs, adopted children will not be as good from their point of view. However, it is another thing to say that this preference is important. One can readily see how the absence of love, or of children, or of spiritual enrichment, or of a sense of family, or of pride in one's accomplishments, would impoverish one's life; but not the absence of biological maternity or paternity. None of the reasons which show that it might be rational for certain people to prefer their biologically own children will support the claim that it is important for them to have their biologically own children. If people were to say that their lives had been substantially impoverished because their children did not share the general facial features of their grandparents, great aunts, and fourth cousins, features which they had rationally desired the children to have, we would hardly know what to make of such an exorbitant claim. Nor would we know what to say of someone who insisted that by being deprived of the ability to create children, he or she had been deprived of something deeply meaningful and spiritually important, similar to the opportunity to worship God or to enter into intimate relationship with other people. If something cannot be important unless there is a reason why it is important, and if we will not accept as important whatever it pleases someone to say is important but only something whose absence would cause life to be impoverished or substantially less satisfactory, then it is difficult to understand how anyone could support the assertion that having one's biologically own children is important.

III. PROCREATION AND THE RIGHT TO PRIVACY

To some people this discussion will seem only marginally relevant to the question of sterilization. The question of 'importance' will seem to them not important, and the consequence that they may suspect lurks behind the conclusion that procreation is not important—namely, that the ability to procreate may be readily curtailed in favor of something that *is* important (such as population control) —they will find simply pernicious. Even if sterilization does not interfere with anything important, they will say, it does affect an area of life that is so personal, so much a matter of the private life of the individual, that it ought not to be imposed on anyone. Procreation, on this view, is a "basic human right."[3] Because of the intimate personal nature of reproduction, one enjoys one's natural powers of reproduction as a matter of right; no additional reasons are necessary to justify the enjoyment of this power, nor can any be as strong. If a person has a right to something, then whether or not it is rational, or even important, for a person to wish to use or have that thing is not to the point; to have a right is to be entitled to do or have a certain thing, whether or not the doing or having of it is rational or irrational, wise or foolish, important or trivial. (Of course it may be unwise or even morally wrong to exercise one's rights, but that is another matter.)

That the power to reproduce is protected as part of the right to privacy is a view that has been suggested by the U.S. Supreme Court. As early as 1942, in a decision overturning a state statute authorizing the sterilization of certain convicts, Justice Douglas declared: "We are dealing here with legislation which involves one of the basic civil rights of man. Marriage and procreation are fundamental to the very existence and survival of the race."[4] This language is typical of the way in which the Supreme Court in later years came to approach the problem of reproductive freedom. The Supreme Court has held in the contracep-

---

[3]The term is used in an article by Gloria S. Neuwirth, Phyllis A. Heisler, and Kenneth S. Goldrich, "Capacity, Competence, Consent: Voluntary Sterilization of the Mentally Retarded", *Columbia Human Rights Law Rev.* **6**, 454 (1974–75).

[4]*Skinner vs Oklahoma*, 316 U.S. 535 (1942) at 541.

tion and abortion[5] cases that the constitution, through the 1st, 4th, 5th, 9th, and 14th amendments, recognizes a fundamental right to privacy that encompasses the right to marry, engage in sexual intimacy, have a family, and decide for oneself whether or not one wishes to bear or not to bear children. Hence, the Court has concluded that the state may not prohibit the dispensing of birth control devices, the state may not prohibit abortion, and the state may not in other ways restrict maternity or paternity, for instance, by firing pregnant employees. The Court has said that there is a right "to be free from unwarranted governmental intrusion into matters so fundamentally affecting a person as the decision whether to bear or beget a child."[6] And the Court has also stated "that the right of procreation is among the rights of personal privacy protected under the constitution."[7] These pronouncements by the Court suggest the direction in which U.S. Constitutional law is moving.

But are these dicta by the Court well-founded? The Court is clearly overstating the matter when it says that procreation "fundamentally affects the person": perhaps it is confusing having children with creating children. To determine whether the right to create children is or should be part of the right to privacy, we must examine this latter right. Evidently when we say that something is or should be private, we mean one of three things. Either we mean that no one but the person concerned has the right to make it public, i.e., known to others (this I call privacy-1); or we mean that no one but the person concerned has the right to make decisions about it (privacy-2); or we mean that no one but the person has the right to use it or do various things with it (privacy-3). Privacy-3 might perhaps be thought of as a variant of privacy-2, but there is sufficient difference between the right of privacy that protects a person's right to marry whom he or she chooses (privacy-2) and that which protects one's use of one's name, or one's use of one's body (privacy-3) to warrant a distinction. (Some matters, for instance abortion, are private in both senses.) In the first sense, privacy-1, the right to privacy is a right to confidentiality. We demand privacy in this sense for things we consider shameful, embarrassing, or ridiculous; or for things we consider very close to us, such as the identity of our sexual partners or the size of our income; or for things that might be revealed to others only with risk; or for anything about which we

[5]*Griswold vs Conn.*, 381 U.S. 479 (1965); *Roe vs Wade*, 410 U.S. 113 (1973); *Doe vs Bolton*, 410 U.S. 179 (1973). There is now a line of cases dealing with privacy in general and reproductive freedom.
[6]*Eisenstadt vs Baird*, 405 U.S. 438 (1972) at 453.
[7]*Rodriguez vs San Antonio Independent School*, 411 U.S. 1 (1973) at 34, note 76.

think the public has no legitimate interest, whether it be important, "personal" or not (for example, the confidentiality of our mail, which we think is our private business, whether or not it contains any information we wish to keep secret). We consider our privacy in this sense invaded when stories about us appear in the newspaper, or when our words are broadcast over microphones we believed to be turned off. There are also things we prefer to do in private, that is, unobserved, although we may have no special dislike if other people know we do them: sexual intercourse and using the toilet are obvious examples, but very often we prefer to sit home and relax, doing nothing in particular, enjoying the freedom from observation provided by walls, fences, and shrubbery. The privacy protected in U.S. Constitutional law by the Fourth Amendment's restrictions on governmental searches and seizures is privacy in this sense.

In general, privacy-1 embraces the protection of information, the freedom from being watched, the protection of certain areas of confidentiality such as mail and conversations. Privacy in the second sense concerns the right to make certain decisions affecting our life. When we say that such things as the use of contraception, the decision to have an abortion, and the right to read literature of our choice are matters of privacy, we do not mean that information about them is protected (although this might also be the case; a woman might not want others to know that she has had an abortion, for example), but that the right to choose concerning them is not to be interfered with. It is in this sense of privacy that we say that certain important decisions, such as whom to marry, are private matters (if you are a member of a royal family, however, this decision is not a private matter; the public or its representatives has a right to participate in decisions of great public moment). Privacy-2 protects certain things that are important; or about which the individual is in a better position to decide than anyone else; or which concern matters of spiritual or intellectual significance, such as one's choice of religion one's moral opinions, or one's political affiliation; or which concern matters of individual self-expression, such as dress, hair style, or choice of friends; or which concern innocent pleasures of no particular significance to the world at large, such as one's choice of recreation, or what one eats, or what movies one sees. It is evidently this sense of privacy that is meant when the Supreme Court says that the right of privacy protects procreation.

Finally, privacy-3 protects the use to which we put certain things, such as our names, our pictures, and our bodies, which are in some intimate sense peculiarly personal or "our own." In this sense privacy is invaded by the unauthorized use of our name in commercial advertisements, or by intrusions into our bodies in the course of police investigations. The Fifth Amendment's proscription of compulsory self-incrimination protects privacy in this sense; it is not that the government has no right to find out certain information (that would be a violation of privacy-1), but that it may not require individuals to provide this information against themselves. Invasions of the body in the search for evidence (pumping the stomach to discover drugs, e.g.) also arguably violate peoples' rights to use their bodies as they choose.[8]

Despite their very great importance, privacy-2 and privacy-3 are often overlooked by philosophers,[9] or, if recognized, confused with something else. Elizabeth Beardsley, for example, who explicitly recognizes this second sense, says that privacy in this sense is violated when "one person Y restricts the power of another person X to determine for himself whether or not he will perform an act A or undergo an experience E."[10] But this confuses privacy-2 with liberty: to "restrict the power of another person" is to violate that person's liberty, not his privacy. This is a tempting confusion, since privacy-2 is closely connected with liberty: Berlin for example sometimes thinks of his "negative liberty," i.e., the absence of coercion, as establishing an area of privacy,[11] although the connection between negative liberty and privacy is strictly contingent: there could be negative liberty in the public realm exclusively [if, for example, the only liberties allowed were the classical

---

[8]In Justice Douglas' opinion, such intrusions into the body violate the Fifth Amendment prohibition of self-incrimination. See his concurring opinion in *Rochin vs California*, 342 U.S. 165 (1952): ". . . capsules taken from [an accused's] stomach . . . without his consent . . . are inadmissible because of the command of the Fifth Amendment." In *Schmerber vs California*, 384 U.S. 757 (1966), Douglas said of intrusions of this kind: "no clearer invasion of the right of privacy can be imagined."

[9]For instance, all of the articles in the recent series on privacy in *Phil. Public Affairs* **4** (4) (1975); **6** (1) (1976); and **6** (3) (1977)—articles by Thomson, Scanlon, Rachels, and Reiman; and the letter by Fried—appear to restrict privacy to the first sense. Thomson's examples of violations of privacy include such things as looking at people through the walls of their house with an X-ray machine and peeping at their pornographic pictures; Scanlon talks about "zones" or "territories" to be protected against "intrusions" which are "observations of our bodies, our behavior or our interactions with other people" (Vol. **4**, p. 315); Reiman (discussing Rachels) says "if there is a unique interest to be protected by the right(s) to privacy, it must be an interest simply in being able to limit other people's observation of us or access to information about us" (Vol. **6**, p. 30). To say this is simply to ignore privacy-2 and privacy-3.

[10]Elizabeth Beardsley, "Privacy: Autonomy and Selective Disclosure", in *Privacy*, J. Roland Pennock and John W. Chapman, eds. Nomos XIII, New York: Lieber-Atherton, 1971, p. 56.

[11]Isaiah Berlin, *Two Concepts of Liberty*, Oxford, Oxford University Press, 1958, pp. 10–11.

political (hence, public) liberties of free speech and free press].[12]

The confusion between liberty and privacy-2 perhaps arises because privacy-2 can be used either evaluatively or quasi-descriptively. We use it evaluatively when we say that certain areas of life should be private: to say this is to say that people ought to be free from interference in these areas. But we also say that these areas *are* private, and here we use the term quasi-descriptively: we are saying that because these areas of life share a certain feature (privacy), they ought to be free of outside interference. Hence *(P)* that something is a private matter, entails *(Q)* that people ought to be free concerning it; but *(P)* cannot be analyzed as equivalent to *(Q)*. *To say* that it is private is not *to say* that people should be free with respect to it.

If privacy-2 simply were the freedom to do certain things, then we might think that we have explained the importance of privacy-2 simply in terms of the general presupposition in favor of liberty. But to do so overlooks that the importance of liberty in the areas of life that we consider private is simply that we do consider them private: we think that people ought to be free to do certain things just because these things are their private business. You cannot therefore explain privacy by liberty; the explanation seems to work the other way around.[13]

---

[12]Berlin has been criticized for this confusion in the following words: ". . . Berlin often uses the term 'privacy' while discussing negative liberty. . . . Clearly, however, liberty and privacy ought not to be equated . . . prisoners placed in solitary confinement, nevertheless have plenty of privacy." From William A. Parent, "Some Recent Work on the Concept of Liberty", *Amer. Phil. Quart.* **11**, 150 (1974). The trouble with Parent's counterexample is that he confuses two senses of privacy: the prisoner has privacy in the first sense (unless being spied on), but this sense is consistent with lack of liberty. In equating negative liberty and privacy, Berlin is clearly referring to privacy in the second sense. Nevertheless, the example does indicate a difference between liberty and privacy-2. Although there are certain matters normally considered one's private business that the prisoner is not free to do (for instance, not being free to play golf whenever one so desires), we would not consider this lack of freedom a violation of the prisoner's privacy. Evidently privacy-2 is a corollary to the more general notion of freedom: if a person is free, then should the choice to engage in sexual intercourse be made, it would be a violation of that individual's privacy to prohibit the use of contraception. But it is not a violation of privacy to put the prisoner in solitary confinement in which there is no freedom to engage in sexual intercourse or to use contraceptives. To violate a person's privacy-2 we must interfere with the specific private act: interfering with that individual's freedom generally, so that certain specific private acts cannot be performed, is not a violation of privacy. The true argument against equating negative liberty with privacy-2 is that an interference with liberty is not necessarily an interference with privacy. If you prevent me from publishing a criticism of the government in my newspaper, you have interfered with my liberty but not with my privacy.

[13]A very important privacy-2-making characteristic, at least in my opinion, is whether the individual is in a better position than anyone else to make the best decision.

We must not understand privacy-2 in too Millean a fashion as "self-regarding" actions of concern to no one but the person doing them. Although one's religious opinions or style of dress may be self-regarding in the Millean sense, it is not true that such private matters as whether to raise a family and what number of children to have, affect only the people making the decision: had my parents, for example, decided to have other children besides me, my life would perhaps have been quite different from what it has actually been. Nevertheless, decisions about how many children to have are considered almost quintessentially private in this sense of private. Clearly many private decisions such as those concerning friends and family affect other people. Procreation, which is our interest here, is surely not self-regarding.

Our problem is whether procreation is a legitimate part of privacy-2. Although we do not have a general theory of privacy-2, we can say that private matters protected by this right seem to fall into six rough categories. (These categories are not necessary or sufficient conditions: they are reasons why something might be considered private.) Something is a private matter and presumably entitled to be protected by the right to privacy if it is of special importance to the individual; if it is of a peculiarly intimate nature, i.e., very close to the individual's self-conception; if the individual is in the best position to make decisions concerning it; if it is a question of spiritual, moral, political, or religious import; if it is a matter of self-expression; or if it is an innocent pleasure of no importance to the world at large. Clearly such things as marriage, sexual preference, style of dress, religious practice, and the like fall into one or many of these rough categories. And just as clearly, or so it seems to me, procreation does not.

I have already argued that procreation is not important, at least where adoption makes possible control over the size of one's family. Clearly the right to procreate is not in the same class of importance with

---

This seems to be one reason why we allow people to decide what careers to enter. We could conceivably have a more regimented system, where people are assigned careers based on society's projected needs for certain kinds of workers, matched to the talents and interests of individuals as established by aptitude and psychological tests. One reason we do not do this is that we think a person generally knows better than any test can establish what career he or she wants to pursue and even what he or she will be good at, in part because we think that people will do better what they want to do and are happy at doing. We feed people information about what is needed, and use test results to guide their intelligent choices, but the right to decide is left to them. In doing this we believe we will get a better match of person to job, overall, than by any other method. This general rule gives rise to a universal right. When, however, we are less confident about the individual's ability to make the best decision, we do not conceive of any right to choose: required courses in universities are a good example.

the right to choose one's mate, to determine the number of one's children and hence to use contraceptives, to enjoy the sexual pleasure of one's choice (and hence to have access to pornography), or other aspects of family and sexual life. We have acknowledged that, for certain people, there might be something spiritually or even metaphysically significant about procreating, carrying, and bearing children, so that the right to procreate might be thought of as part of the right to spiritual or religious freedom. Yet very few people would argue that procreation is so spiritually important that they want to enjoy it over and over again: people who have large families presumably like children, not procreation. Hence, inclusion of procreation in the right to privacy based on analogy with matters of religious or spiritual meaning seems tenuous. Nor can it be plausibly held that procreation is an innocent pleasure, such as eating or sex, which is therefore the individual's private business; nor does it seem a matter of self-expression, as do dress and hair style. Procreation is sometimes thought of as an intimate personal matter, close to one's conception of one's self. We have acknowledged that pride in creating children might be a legitimate value, but since procreation is a biological function, not a human activity, it is irrational to make this pride central to one's conception of one's self. People who clearly keep in mind the difference between procreating and having children, the difference between sharing mutual love and sexual pleasure and causing conception, and the essentially biological nature of procreation as opposed to truly human activities, are unlikely (one would hope!) to think of themselves as in some important way reproducers. Reflection can change one's conception of oneself. Matters that may at first seem very close and personal begin to fade as reflection puts them in their true light; this I believe to be the case with procreation. I conclude,then, that contrary to the opinion of the U.S. Supreme Court and others, procreation does not seem to be a matter of personal privacy, and there seems no reason why the ability to procreate should be protected by a right to privacy-2.

Nevertheless this is not the end of the story. Although the ability to procreate should not be protected by a right to privacy-2, it does not follow that involuntary sterilization does not violate the right to privacy. Given present technology, involuntary sterilization violates the right to privacy in the third sense. This is so because sterilization is a medical procedure that involves an intrusion into the body, and can therefore be done upon unwilling subjects only by coercion, physical force, deception, or manipulation. Intrusions into the body, when effected by these means, are acknowledged to be violations of privacy -3. (But not always, at least not in U.S. Constitutional law, which allows, e.g., the involuntary taking of a blood sample to test for alcohol content in certain situations.)[14] Should technology develop procedures by which people could be sterilized without such gross intrusions, for example by radiation administered during sleep, the case against sterilization as an invasion of privacy-3 would be weaker, though doubtless many people would feel that irradiation, or some similar procedure, is a sufficient intrusion to count as such a violation. (But suppose we imagine a world in which people could be sterilized directly by the command of some psychokinetic superhuman being?) Involuntary sterilization also potentially violates the right to privacy-1 in that, at least if done on any large scale, it requires the keeping of extensive records. And although the collection of personal information is not itself violative of the right to privacy, such records are notoriously open to abuse and difficult to control, and hence should not be maintained unless strictly necessary.

I conclude from all this, then, that maintaining the natural ability to procreate is not important in itself and is not protected by the right to privacy. However, the procedure by which this ability is destroyed does involve actual and potential violations of the right to privacy in the first and third senses. Since involuntary sterilization involves the abridgement of a right, very strong justifications are needed if involuntary sterilization is to be justified.

IV. STERILIZATION AND POSSIBLE PEOPLE

In the remainder of this paper, I want to discuss one justification sometimes proposed for sterilizing certain people. This justification has to do with the fact that some people carry genetic defects that might be harmful to their offspring. Now this fact raises many important ethical problems, involving questions from abortion and infanticide to the allocation of medical costs between society and patients. None of these problems can be discussed here. The question I do wish to discuss is a conceptual question: does it make sense to say that sterilization can be justified to protect the possible victims of genetic defects?

A similar problem arises with abortion. Sometimes it is said that we are justified in performing an abortion (on a thalidomide baby, say) for the fetus' own sake, or in the interest of the fetus. Others, however, take the position that abortion can never be in the interest of the fetus, since nothing can be better off dead, and it simply is senseless to say

---

[14] *Schmerber vs California,* 384 U.S. 757 (1966).

that we can protect something by killing it.[15] Whether something is better off dead is a normative question; there might be conditions under which someone's life were so horrible that death would be the only protection from suffering. Suppose, however, we decided that abortion for the fetus' sake was never justifiable. Recognizing that sterilization does not involve the taking of a life, we might conclude that in order to protect possible victims from genetic disease, sterilization of the potential parents (when less drastic methods of birth control could not be expected to work) might be justified. We might take the position, say, that when two carriers of some serious disease wish to marry each other, the price they will have to pay must be the forfeiture of their right to have their own biological children (we can mitigate this price by putting them high on the priority list to receive adopted children). We could justify this course by saying that the parents in transmitting the disease to their children are harming them, and society has the right to protect innocent people from harm. But at this point puzzles begin to appear. If it seems strange to say that something can be protected by killing it, it will seem even stranger to say that it can be protected by keeping it from being conceived, i.e., by preventing it from coming into existence in the first place. In the case of abortion, the question seemed clearly normative: we are asked to balance whatever value the life of the person whom the fetus will become will have against the suffering it will have to endure. But if the question about preventing conception is a normative question, which question is it? Is it the same normative question as the question about abortion? Or is there a difference in value between being conceived and then aborted, and not being conceived at all? If there is a difference in value, it would seem that it is better to be conceived and aborted: one has gotten closer to human life, in a way, and has at least had some existence, if that is in itself valuable. But if so, shouldn't abortion seem a preferable course to sterilization, since abortion at least allows the person we are protecting to have existed for a while?

But apart from these normative puzzles, there are conceptual puzzles. When we abort someone, we know to whom it is that we are doing it. Fetuses are identifiable individuals who have rights and interests, and can be harmed and benefited. But when we prevent someone from being conceived, we are preventing that individual from even coming into existence, and then it is not clear who it is we are benefiting (or harming). A harm or a benefit is conferred on something, but in this case

---

[15]See Paul F. Camenisch, "Abortion: For the Fetus' Own Sake?" *Hastings Center Report* **6(2)**, 38 (1976).; and letters, *Hastings Center Report* **6**, pp. 4ff (August, 1976).

there is no 'something' on which to confer 'anything.' Even if we were willing to grant, as a normative matter, that something could conceivably be benefited by being prevented from ever coming into existence, how can we identify the something being benefited when that benefit consists in preventing its coming into existence?

Let us note two things. First, we are not saying in general that it is impossible to benefit (or harm) the unconceived or nonexistent.[16] We can harm or benefit future generations, for example, by wasting or conserving resources, and the members of future generations are presently unconceived and nonexistent. Second, we are not saying that there is no sense in which nonexistent things or people can be identified. As Hare has pointed out,[17] we can identify nonexistent people by describing them, though of course, we cannot say who they are (i.e., which real existing persons they are; we cannot identify them *as* someone). Just as we can identify the person who will next occupy this carrel in the library by that description, though we cannot identify the person as any particular individual, so we can identify a nonexistent person as that individual who will be born if these two people begin coitus in five minutes; and we can distinguish that person from the one who would have been born had they started five minutes ago. But although some nonexistent people can be the subjects of harms and benefits, and all nonexistent people can be described at least minimally, not all nonexistent people can be identified as particular people. The person who will occupy the carrel is some particular person or other, though we cannot say who. But the person who would have been born is no particular person at all. When we are talking about future generations, we are talking about people who will, or at any rate might, come into existence in the future: they will be real particular people, should they come into existence, even though they do not happen to be particular people just now. But people who have been prevented from coming into existence through the sterilization of potential parents could not be real particular people. We claim that we are benefiting these people by making it impossible for them to exist, i.e., by making *them* impossible. And an impossible person is not—could not be—any one in particular. If, to use another of Hare's examples, radiation were to sterilize everyone so that there are no future generations, then whom would we have benefited

---

[16]For a discussion of many related problems see Michael Bayles, "Harm to the Unconceived," *Phil. Public Affairs* **5**, 292–304 (1976).

[17]See R. M. Hare, "Abortion and the Golden Rule," *Phil. Public Affairs* **4**, 219ff (1975). Bayles' argument against Hare on this point (note 16 *supra*, pp. 298–300) does not seem convincing.

by carefully conserving our resources? It seems clear that the answer is: no one. We consider benefiting nonexistent people because we think they might become existent, i.e., real particular people, some day. But the person who did not come into existence is nobody in particular. And it is hard to see how a benefit can be conferred on nobody in particular.

And there is another point. The number of impossible people associated with any potential parent is enormous: approaching infinity. For at any given moment, the person might have had intercourse leading to the production of a new person, and that person would have been a different person from the one who might have been produced at any other moment. All of these possible people, however, are rendered impossible by the very fact that the person did not have intercourse at most of these moments. If, however, to bring someone into existence is to benefit that person, and if people who were not brought into existence are proper subjects of benefits and harms, then every person is responsible for failing to benefit countless persons throughout his or her lifetime. This conclusion seems bizarre at best, but it also seems unavoidable if we are to maintain that persons who were not conceived because their parents were sterilized have been benefited.

There is, finally, one last point. It is sometimes said that if potential parents were sterilized to prevent the birth of children with genetic defects, the world would inevitably lose some valuable people. This is a very bad argument. Whatever plausibility it has arises from knowledge gained only after the fact: we can point to someone whose parents would have been sterilized and say that that person would not have come into the world. But any contraception entails the loss of a possibly valuable person; we do not feel a loss, however, because the person has never been there to lose. The "but we would have lost X" argument can only be an argument for the greatest feasible quantity of reproduction: if a couple who decide to stop at n children had only gone on to $n + 1$, who knows who it might have been! Of course, the same bad argument is sometimes made against abortion ("If my parents had practiced abortion, I might not be here now"), for which it is equally invalid.

ACKNOWLEDGEMENTS

I wish to acknowledge a debt to Michael Pritchard, Michael Bayles, Elaine Wilson, and Emily Michael, all of whom read the penultimate version of this paper and were generous with their comments.

Comments on "Sterilization, Privacy, and the Value of Reproduction"

Michael D. Buckner

Philosophy and Medicine Program, New York University, New York, New York

Joseph Ellin has written a provocative paper on the matter of procreation and sterilization. Although he concludes that involuntary sterilization violates a basic right to privacy, not many will be satisfied with his argument that this is unrelated to a right to procreate, and that the latter interest in procreation is not an "important" interest to individuals. As he notes, having robbed procreation of both the strong status of a right, and the weaker status of an important interest, the road is cleared for an unfavorable comparison with truly important, and most likely public interests. Although the cleanup tasks are left for another occasion, Ellin himself assumes that from the perspective of public policy procreation is quite important, and thus as a matter of public policy, it will require legislation and control. The remaining safeguard against control is based upon the fact that effecting control requires an "intrusion into the body." Therefore it is a matter of violating that sense of privacy which protects the use of our body. He seems to think that to the degree that actual physical intrusion is not necessary, to that degree privacy will not be violated. Therefore, since there is no other reason for banning the effecting of control through sterilization, such sterilization would be legitimate.

I find this to be an odd bit of reasoning. First, it seems to focus upon an act of bodily invasion as definitive of "sterilization," at least with respect to its morality. It strikes me that sterilization is a good deal more than the momentary utilization of a person's body for purposes against that person's interests (be they ever so "unimportant"). Actually, Ellin has not argued the stronger thesis that it is involuntary sterilization which is a moral and legal violation, but that the use of the person's body is the violation. He is discussing the means to effect an end, the means being an "invasion" and the end being sterilization. The means are illegitimate; there is no separate argument supporting the proscription of the end. When we realize this, we realize that the net effect of Ellin's paper does not truly support our intuitions about sterilization and procreation, but leaves them exposed to the vagaries of public policy formation. Indeed, the example of the relevant sense of privacy that he maintains protects us against involuntary sterilization is instructive. He notes that although it is a right of privacy not to have our bodies manipulated against our will, it is not an inviolable right. He cites the case of compulsory blood tests to determine the alcoholic content of drivers suspected of drunken driving. He does not pursue this further, but the conclusion is obvious. Since violating a person's right to privacy is justified by consideration of the public interest in matters involving automobile traffic and its attendent fatalities, then it should be apparent that procreation (and the consequent population and economic problems) are of even greater public concern. Therefore, they would justify even more strongly the abrogation of the individual's right to privacy in order to effect sterilization. The matter is settled on Ellin's account by reference to matters of public interest, since there are no morally relevant features specific to procreation and sterilization themselves.

The above argument appears in the concluding paragraphs of his analysis. I have begun with it because I think that it represents a fundamental flaw in his account. He presents a number of arguments about sterilization and procreation, but fails to analyze these concepts in sufficient depth. His consideration of the applicability of the right to privacy is vitiated by this failure. In the remarks that I have cited above, we are left with the notion that the elimination of the capacity to procreate requires an invasive process of sterilization, and that the bodily invasion is solely a matter of the sterilization technique. Unless he has in mind an unsubstantiated capacity to procreate, I totally fail to see how any technique of sterilization can effect its end without involving a violation of my body. My capacity to procreate is based upon the functioning structure of my body, and to eliminate that capacity, to achieve sterilization, requires the restructuring of that structural–functional system which is my body–myself. The analogy with blood sampling for alcohol is now revealed to be defective. In that case, the invasion is simply to remove a trivial component of the fluid support system in a nonconsequential quantity. The issue is over using a person's body for a few moments, and not at all over restructuring it. However, in the case of sterilization we are not merely protected by a sense of privacy that involves merely this sort of momentary and inconsequential usage. It is not simply a matter of usage, but one of *identity* per se. This comes closest to Ellin's sense of privacy$_2$. The significance of the point is that if procreation, in the context of involuntary sterilization, is protected by a sense of privacy different from that governing mere usage of a person's body, then Ellin's conclusion that this latter sense of privacy is the sole one of relevance is false. If his conclusion is false, then his argument is not sound, even if it is valid. Indeed, I think that his argument may be valid, but only on an interpretation which renders his premises unsound.

Ellin states that the purpose of the earlier arguments in the paper is to establish two claims mentioned at the beginning of this critique, *viz.*, that procreation is not "particularly important," and that the capacity for procreation is not protected by those aspects of the right to privacy that protect marriage and family life. He argues for the latter point by distinguishing three senses of privacy, the third of which we have already discussed. He considers that marriage and family life are protected by the other senses, not the third. We have already discussed his arguments for assuming the relevance of the third sense to sterilization.

Let us scrutinize these opening claims. In his introductory remarks, his statements are about the capacity to procreate, not merely sterilization. We have seen that the relationship between these two requires explicit analysis, and the result upsets his argument. The same difficulty occurs at the outset insofar as the various senses of procreation are not distinguished. "Procreation" may be seen to have three distinct senses: a capacity of individual humans possessed by their structural–functional constitution, i.e., their identity; a process of events that brings about the development and birth of a new individual human; and the actual occurrence of such development and birth. These distinctions are of some moment for both individuals and for public policy, although we cannot pursue such issues here. When Ellin states that "nothing can be more important to the world at large than the number of children people have . . . ," he is discussing the third sense, and not the capacity per se. Thus, such a point is simply not relevant to arguments about the inviolability of the capacity. Sterilization is essentially a matter of the capacity sense, and therefore arguments for sterilization must take into account interests, public or private, pertaining to the capacity itself. These matters are conflated in the paper.

Ellin's aside, in which he distinguishes between privacy and liberty, can be brought to bear on these distinctions. It is the capacity to procreate that may well be protected by a right to privacy. But it is the liberty to employ this capacity and realize its ends that might be subject to matters of public interest. Of course, restricting a person's liberty requires serious justification, and this presumes that the area of behavior is of some importance. This brings us to the matter of 'importance.'

Although Ellin claims that questions of importance and of the applicability of the privacy concept are separate issues, he actually bases his most important arguments against the latter upon the former. He divides the concept of privacy-3 into five categories, and argues that procreation is not covered by this concept since it is not covered by any of its components. An argument by exclusion. Unfortunately, the exclusion from one of the five categories, *viz.,* when something "is of a peculiarly intimate nature, i.e., very close to the individual's self-conception," is based upon an appeal against its importance. Thus, if we deny the soundness of the assumption about importance, then we will block the exclusion from the second category of privacy, and therein block that argument as well. Since I have already asserted that the capacity to procreate is indeed a matter of human identity, it follows that I cannot agree with his exclusion of it from this category of privacy. His other categories were the following: if something is of special importance to the individual; if it is a question of spiritual, moral, political, or religious import; if the individual is in the best position to make decisions concerning it; if it is a matter of self-expression; or if it is an innocent pleasure of no importance to the world at large. (We have already cited the category of relevance to self-concept.) I find that most of Ellin's exclusions are merely matters of fiat. There are a few commentaries. For example on religious and spiritual significance he asserts that few people would "want to enjoy it over and over again." I cannot comprehend the relevance of repetition to the question of the existence of spiritual significance. It certainly does not prove that the analogy is "tenuous."

As for the matter of self-concept, he claims that "we have already rejected this belief as irrational." The rejection occurred in his arguments against the importance of procreation. Since we are discussing the capacity to procreate, his claim is that this capacity is not important to our concept of self. He admits that some believe it to be important, but claims that this belief is confused. They are confusing such matters as love-making, sexual intimacy, and so on with procreation. Furthermore, without such confusion we would realize that procreation is a mere biological activity, and therefore "not a human function." In his own discussion he equivocates between various senses of procreation. The presumed topic was to be the capacity to procreate, but the points pressed during the analysis pertain to the process and the activity depending upon context. He states at one point that "there is no such thing as the activity of procreation; what there is is sexual intercourse, which may be entered into with the hope and expectation that it will result in procreation." Since he speaks of merely biological activities, he cannot mean that procreation is not an activity simpliciter. He must mean that it is not an action, i.e., a course of events controlled by human volition. But of course this is a straw man, since only the capacity to procreate is the topic.

In the current sense procreation is a success term: it refers to a certain outcome of human activity. It is an activity requiring the remarkable conjunction of a variety of events and the realization of many individual potentialities; an activity which can be subject to voluntary initiation and acts of support. Without this initiation, and without this context of volitional support, procreation would not occur. Whereas meitosis is a purely biological activity, human procreation is most certainly more than a mere biological activity.

Ellin's remarks reveal a peculiarly academic and intellectual bias towards certain sorts of human activity, *viz.,* those activities subject to the maximum degree of human mental control: ones in which all phases, initiation, continuation, manipulation, and completion are subject to the highest mental capacities, and for which the minimum physical involvement is necessary. Too great a physical involvement would contaminate the human achievement with mere biophysical processes. The correlative bias is against anyone who takes pride in achievements relying on such nonvolitional processes.

> Since impregnating or being impregnated are more akin to biological functions than to human accomplishments—more like digestion than like hitting a home run—it may be questioned whether this pride is not misplaced.

At this stage of his argument he notes that although he can question the importance of such values, he cannot challenge their rationality, since when a preference is "something founded in a fundamental value, we cannot, by our criterion, say the preference is irrational." Unfortunately, when he refers to this argument in support of the exclusion from the self-concept, he asserts that the *irrationality* of the claim has been established.

Beyond the distinction between rationality and importance, the gist of his remarks is that a concept of identity should not be based upon mere "biological functions," but fundamentally upon human accomplishments. We see that his position is normative: these functions *should not* constitute our self-concept. And if they do enter into that self-concept, then they *should not* be taken as "important." But what makes something a matter of importance? He answers that it is determined by whether its absence would "impoverish one's life." One needs reasons to support a claim of impoverishment. But what sorts of reasons? Ellin does not provide an account of the nature of such reasons, other than the ones that would follow from his value commitment toward human accomplishment. Of course, we all hold such accomplishment as an extremely important feature distinguishing our species from others, and probably one individual from another. Yet, is it true that we therein eliminate the value to our self-concept of biological functions? The biological nature of the process simply cannot serve as a sign of dispensibility. Even digestion, the alleged paradigm of nonaccomplishment, is essential to our lives. Its absence would impoverish them in a rather significant mode. Even where the functions of digestion could be established in alternate ways, a person comes to appreciate the unemcumbered mindless capacity which others possess.

Ellin does not provide us with greater clarity on this concept of "impoverishment." I find it to be less adequate than the notion of "importance." He claims that having children is important, and therefore assumes that their absence would be impoverishing. But for what reason? Of course, their presence brings joy, and so on. But is their absence as important as he seems to require to support his notion of impoverishment? Many young couples are choosing to forego children to pursue a certain type of life-style. This seems to suggest that the presence of children is not necessary to the concept of the self. Similar points could be made about the items he seems to assume pass muster. Furthermore, procreation is not precisely like digestion. When we reflect on the nature of the process of procreation, the entire process, we realize that it is a creative activity. It might be defined as the combination of genetic material from two humans of complementary sex to form a genetically complete organism, and the subsequent development of that organism into a human individual capable of self-supporting biological existence. It is clear that I have tried to skirt several issues in this formulation; but it should be noticed that sexual intercourse is not essential (thus Ellin is wrong to say that there is no procreation, but merely sexual intercourse), and neither is female labor nor natural childbirth crucial. It may be the case that individuals who have this capacity can live fulfilling lives even though they choose not to utilize it, and that other individuals who lack this capacity will also be capable of living fulfilling lives, and therein all such individuals will not lead impoverished existences. But we have seen that the same applies to *all* other objects, activities, etc. that Ellin, and the Supreme Court for that matter, have recognized as matters of privacy. Ellin's identification of these as matters of impoverishment seems unsupportable without further comment.

Ellin notes that "reflection can change one's conception of oneself." This is most certainly the case. Unfortunately, there is no a priori argument that people' self-conception will move in the direction of relegating all biologically based activities to the realm of the nonimportant. It seems that the more information one has about the procreative process, the more valuable procreation itself becomes, and the less valuable certain nonessential concomitants, e.g., sexual fidelity. Others have commented that with the separation of sexual activity from procreation, sexual activity has lost its "spiritual" significance and has been freed to a more common realm of daily activity. It is interesting to reflect that savages ignorant of the nature of procreation might be talked much more readily into sterilization, since they would see it merely as the manipulation of their body, and might more readily consent to that than to the elimination of their procreative capacity. Whether even more knowledge and reflection will lead us to support Ellin's diminution of the procreative capacity remains to be seen.

Finally, we must note that the structure of the argument has been inverted. We have come to see procreation as constituting one's concept of oneself as a human being. On Ellin's criterion it would qualify for protection as a matter of privacy-3. The argument now turns upon whether it is an important component of our identity. We cannot argue separately about whether it is an important *interest* if we believe that it is a matter of interest because it is a matter of our identity. His separation of the issues does not stand. It is in the realm of rights, whether we explicate this by reference to privacy, or some other fundamental concept of dignity and autonomy, that procreation must be discovered.

The realm of rights is not inviolable. Considerations of importance operate in conflicts of rights and interests. But we have not seen any satisfactory arguments that go beyond exhortations based upon Ellin's intellectualistic value perspective. There are a great many people whose lives are in fact enriched, and whose lives would in fact be significantly impoverished, if they could not procreate. (I am referring here to the knowledge that they have produced their own biological children.) I am not clear that to argue that they should not set such store by these accomplishments, since they are merely biological, is not to argue that they should live according to a totally different set of values that would in fact require them to be different sorts of persons. One would be arguing that people ought to reorder their priorities and live different sorts of lives. If such lives require different sorts of characters effectively to be realized, then one is exhorting people in the most classical evangelical fashion. I take such implications to be unequivocal signs of the realm of the personal and private. Ellin has provided no sound arguments to the contrary, and indeed has indicated by his comments that these issues remain a matter of fundamental value and exhortation. I am gratified to be able to support the Supreme Court in this matter.

His concluding section discusses the logic of the appeal to permit sterilization in order to protect the possible victims of genetic defects. This section includes much discussion about protecting possible people, identifiable future individuals, and impossible people. I find this area, along with much of modal logic, to be filled with a great deal of extremely arbitrary gibberish, notwithstanding the fact that the allegedly most brilliant philosopher of our era has provided us with a definition of truth that clarifies everything. If an individual is truly impossible, I suppose that it makes no sense to protect him/her/it; although, perhaps by the time of this writing, our esteemed logicians may have already proved that one can indeed identify impossible individuals across impossible worlds! As for me, I prefer not to speak of the impossible, even if they are not unspeakable.

10

Reply to Buckner

Joseph Ellin

Department of Philosophy, Western Michigan University, Kalamazoo, Michigan

In "Sterilization, Privacy, and the Value of Reproduction" I argued that the capacity to reproduce is not important, and is not protected by any right to privacy, so that if involuntary sterilization is impermissible, it is not because of any "right to procreate." But I pointed out that sterilization which, given present technology, and perhaps any technology, involves an invasion of the body, does seem to violate the right to privacy in what I distinguished as the third sense of that right. Professor Buckner evidently thinks that interference with procreative capacity is itself a "moral violation." Since he denies my argument is sound, even though conceding that my logic is valid, he presumably holds either that the capacity to procreate is important or that interference with this capacity is a violation of the right to privacy. I cannot see, however, that he has shown either of these possibilities to be true.

My reasons for thinking that the capacity to reproduce is not important were first, that people who believe it is important are very likely confusing reproduction with other matters, for example, sexual intimacy or raising children (which I do think are important); and second, that the loss of the capacity does not seriously impoverish one's life. I conceded that the preference to produce one's own children, rather than adopt, might be rational, but argued that rational preferences are not necessarily important, since something should not be considered important unless its absence impoverishes one's life. (Irrational preferences, in my view, cannot be important, since if there is no good reason to prefer one alternative to another, a life cannot be impoverished when it contains the one rather than the other.)

Evidently Professor Buckner disagrees with my conclusions, but whether he rejects my analysis of importance, or simply thinks it misapplied when used to draw my conclusion about procreation, I cannot tell.

His sole objection about my analysis, so far as I can see, is that I have not explained precisely enough what impoverishment means. Perhaps so, but this hardly vitiates my contention. "Impoverishment", though perhaps unclear, is not meaningless, useless, or even unimportant: I can quite truly assert that if I did not have any children, my life would be less rich than it is now, and to that extent, impoverished. Incredibly, Professor Buckner misunderstands me to say that childless adults necessarily lead impoverished lives, as if I had said that the absence of some things is necessarily impoverishing, whereas all I was saying is that the absence of other things (the capacity to reproduce) is necessarily *not* impoverishing: no one can plausibly claim that his or her life is impoverished just because he or she cannot produce their biologically own children. The reason for saying this is simply that all the riches of life, all the central values of life, whatever they may be, intellectual, voluntaristic, even biological, are as available to the person unable to create his or her biologically own children as to anyone else. Certainly that person cannot have all the experiences of life (childbirth, for example), but no impartial spectator, I believe, would judge that these experiences, however rewarding they may be, are the kinds of experiences whose absence impoverishes a life.

Buckner, on the other hand, thinks that the loss of the capacity to procreate involves a "restructuring" of the body and hence a loss of one's "identity" (or even "human identity"). I quite agree that preservation of a person's identity, whatever that means, is important, and even that a person's identity should be protected by the right to privacy. But sterilization, at least in the male, involves no more restructuring than a tonsilectomy, and poses no more threat to one's identity.

But perhaps what Buckner is getting at in the passages about identity is one's self-concept, which he discusses at length (the difference between identity and self-concept is: your identity is what you are, your self-concept is what you think you are). This pertains to my discussion of the right to privacy. Buckner, as I understand him, thinks that I am mistaken in holding that what I call privacy-2 does not protect one's reproductive capacity, since this capacity is (or could be) part of one's self-concept, and privacy-2 protects people's self-concepts. I claim, however, that this is only true of self-concepts that are rational, and a self-concept in which reproductive capacity is a significant element is not rational. Now this claim cannot be made to go away by condemning it as an "intellectualistic value perspective." Buckner must show either that a self-concept that includes the capacity to reproduce as a major element is rational, or that it is irrelevant whether it is rational or not (i.e., he must argue that all self-concepts should be treated equally). I cannot tell which of these choices he prefers, but possibly he would argue both, though they are not entirely consistent. Most of his paper seems to indicate that he would adopt the first position, though accusing me of "intellectualism" seems to indicate that he adopts the second one as well. To hold, however, that people are entitled to protection against action that damages their ill-grounded self-concepts, is to say in effect that public policy can be legitimately thwarted by merely arbitrary or foolish beliefs. Despite his rejection of intellectualism, Buckner would not hold, I hope, that a macho police-man, whose self-concept depended heavily on his false belief that only men can do police work, would thereby have the right to obstruct entrance of women into the police force. I cannot see why any other irrational self-concept is more entitled to protection.

Evidently, then, Buckner must hold not that all self-concepts should be treated equally, but that it is *not* irrational to rest your self-concept heavily on your ability to procreate. The chief reason I gave for holding that it thus is irrational was that reproduction is more a biological activity like digestion, than a human activity like hitting a home run. Buckner's three-fold analysis of "procreation" does not damage this point in the least. Buckner insists that procreation is a "creative activity." But though he takes me to task for saying that strictly speaking there is no activity of procreation, but only sexual intercourse engaged in with the hope that it will result in procreation, he does not even attempt to explain in what sense procreation is a creative *human* activity. Nonetheless I don't deny that people can take pride in procreating; this was one of the reasons I gave for saying that the preference to procreate might be rational. But the similarity be-tween procreation and digestion, it seems to me, renders it irrational to make this pride central to your concept of yourself, and hence to argue that it should be protected by the right to privacy. To think of one's self in some central way as a breeder is not unlike thinking of one's self as a digestor. There is as much achievement, mastery, or accomplish-ment in the one as in the other.

As to whether my views are offensively academic and intellectual-istic, I would remind Professor Buckner that my example of a human activity was not reading Plato or listening to Bach, but hitting a home run, hardly something requiring "minimum physical involvement." I never thought to deny that there are important human activities that engage capacities other than intellectual ones; evidently it is Buckner's Cartesian bias to suppose that all activities are either intellectual or biological: hence, he imagines that if I deny that biological processes are genuinely human activities, I must suppose that the only genuine human activities are intellectual ones. But my complaint is about people who have irrational views of themselves: I am not saying that only intellectualistic views are rational.

Buckner accuses me of "exhorting people in the most classical evangelical fashion" and asserts that my position is "normative." If this is his way of noting that I would wish that people abandon irrational ideas, especially when these ideas present obstacles to the well-being of other humans, then I plead guilty. Perhaps there is something to be said for the view that our poorly thought-out notions deserve as much protection as our well-thought-out ones; but if there is, Professor Buckner has not said it.

III

THE ROLE OF
THE PHYSICIAN

Hippocrates Lost, A Professional Ethic Regained: Reflections on the Death of the Hippocratic Tradition

Lisa H. Newton

Department of Philosophy, Fairfield University, Fairfield, Connecticut

In the present article, it is my intent to consider the morality of the Hippocratic Oath in the context of a larger discussion of bioethics. That discussion is itself so remarkable, and so contrary to the morality of the Hippocratic Oath, that I shall take its existence as the starting point for my present reflections. The existence of bioethics is remarkable because it presupposes a common ground of dialog between philosophers and physicians, a willingness on the part of both professions to contribute insights from their specialized literature and their diverse experiences to the solution (or at least the clarification) of problems that sprawl across the historic boundaries of either one. And it is contrary to the morality of the Hippocratic Oath that such willingness should be there on the part of the physicians. Those in the field tend to begin all such articles or discussions with a comment on the "recent interest in bioethics": this odd mannerism should call our attention to the profound changes in medicine proper that have made discussions of bioethics at all possible, changes specifically in the very influence of the Hippocratic Oath on those practicing in the field.

In what follows, I will consider briefly the Hippocratic ethic as it comes down to us, focusing on its secretive nature and the closed body

of knowledge that characterized the professional wisdom of the physician. Then I will look at the breakdown of this ethic of exclusive
possession in the modern era, consider some of the consequences of the
breakdown (including the emergence of the possibility of bioethics!),
and suggest a line of development for the future of the professional ethic
of medicine based on these consequences.

I. THE HIPPOCRATIC TRADITION AND THE OBLIGATION OF SECRECY

The Hippocratic corpus presents us with (at least) two writers, one of
whom is not the least interested in ethics. That one, of course, is
Hippocrates himself, a physician of the fifth century BC who wrote
treatises on fractures, on dislocations of joints, and on wounds in the
head, among other subjects.[1] He was primarily a clinician, making
accurate observations of the injuries and diseases in his experience to
the end of more successful treatment; he swore by no gods, propounded
no elevated morality, and is of no further interest to us.[2] The morality
of the Hippocratic Oath, to which this paper is addressed, is the morality of different writers. The writer of the *Oath,* we are told, was probably a Pythagorean, for the *Oath* seems to contain formulae and restrictions that were part of Pythagorean teaching, and obviously not
appropriate to medical practice.[3] The *Oath*'s injunction against the use
of the knife, for example, is clearly incompatible with the advice on
incision in *Wounds in the Head* and Hippocratic practice generally.[4]
The writer of the *Decorum,* from which the major supporting texts for
this section are drawn, seems to have been a teacher of the third century
BC, perhaps influenced by the Stoics, speaking to a group of students
already familiar with the medical profession of their time.[5] These are
the writers who have given direction to the morality of medicine in
ancient times and to the present day; it is that morality that puts the

---

[1]E.T. Withington (trans. and ed.), *Hippocrates,* (Loeb Classical Library), Vol. III,
New York, Putnam 1972.

[2]Dickenson W. Richards, "Hippocrates and History: the Arrogance of Humanism," in R.J. Bulger, ed., *Hippocrates Revisited,* New York, MedCom Press, 1973, pp.
14–29.

[3]Ludwig Edelstein, *The Hippocratic Oath: Text, Translation and Interpretation,*
Baltimore, Johns Hopkins, 1943.

[4]Withington, *op. cit.,* p. 31.

[5]W.H.S. Jones (trans. and ed.), *Hippocrates,* (Loeb Classical Library), Vol. II,
Cambridge, Mass., Harvard University Press, 1923, p. 271.

fundamental obstacles in the way of any cooperation with the likes of philosophers.

Consider the first promise of the Hippocratic Oath:

> I swear . . . to give a share of precepts and oral instruction and all the other learning to my sons and to the sons of him who has instructed me and to pupils who have signed the covenant and have taken an oath according to the medical law, *but to no one else.*[6]

The Pythagoreans were a secretive lot, and the Oath is the oath of a secret society of physicians. Fraternities, secret societies, were familiar solutions to the problem of association in the days of the collapse of the Greek city-state, and professional societies were not uncommon.[7] Their "secrets," not just the skills of their profession, but the religious formulae believed to protect their practice, would be closely guarded for reasons of religion as well as trade-union protectionism. The gods were believed to resent public discussion, and it has always been believed that magical powers were diminished by sharing. But there was another reason why these societies took as their paramount obligation the protection of their secret knowledge. The entire tradition of the cults, from which the societies came, linked an individual's ability to attain knowledge to that person's state of morality and sanctity. As the writer of the *Law* puts it, some centuries after Hippocrates:

> Things however that are holy are revealed only to men who are holy. The profane may not learn them until they have been initiated into the mysteries of science.[8]

Until the gods and the guardians of the cult have accepted a student, that person not only should not, but *cannot* acquire that wisdom "useful for many things . . . , that is, wisdom applied to life," referred to in the *Decorum.*[9] Secrecy vis-à-vis outsiders is therefore, from the perspective of such a fraternity, necessary for general as well as their own safety, for their "wisdom," their medical knowledge, *cannot* be learned by outsiders, and its counterfeit might easily be abused. So it is to holy secrecy that the physician is pledged upon entering the profession: "In purity and holiness I will guard my life and my art."[10] And it must be the secrets to be guarded that are referred to in the last sentences of the *Decorum:*

---

[6]Edelstein, *op. cit.,* p. 3, (emphasis supplied).

[7]Jones, *op. cit.,* p. 273.

[8]*ibid.,* p. 265.

[9]*ibid.,* p. 279.

[10]Edelstein, *op. cit.,* p. 3.

Such being the things that make for good reputation and decorum, in wisdom in medicine, and in the arts generally, the physician must mark off the parts about which I have spoken, wrap himself round always with the other, watch it and keep it, perform it and pass it on. For things that are glorious are closely guarded among all men.[11]

The whole practice of medicine, then, for these physicians three or four generations removed from Hippocrates, was bound up with secret knowledge, concealment, and disguise. The authority of the physician rested on his possession of the secrets and experience in practicing according to their instructions. In this light we can read the *Decorum*'s suggestions for the physician's dealings with a patient:

Perform all this calmly and adroitly, concealing most things from the patient while you are attending to him. Give necessary orders with cheerfulness and serenity, turning his attention away from what is being done to him; sometimes reprove sharply and emphatically, and sometimes comfort with solicitude and attention, revealing nothing of the patient's future or present condition. For many patients through this cause have taken a turn for the worse, I mean by the declaration I have mentioned of what is present, or by a forecast of what is to come.[12]

The last sentence purports to justify the advice of the two previous sentences. It is somewhat lame; but such justification, as we have seen, does not have to carry too much weight. The injunction against telling the patient the truth about her or his condition need not derive from a utilitarian calculation of the patient's well-being; it is a simple part and extension of the general injunction not to tell *any* nonphysician about *anything* medical. The secretiveness of the physician is not "paternalistic," it is religious, and has nothing to do with the patient's health.

The physician of the Hippocratic tradition, then—and the *Oath, Law,* and *Decorum* are still collected under the name of Hippocrates, even though Hippocrates of Cos had nothing to do with these or any works of this nature—claims possession of a secret "wisdom," a wisdom useful in human affairs, and takes that wisdom to be the essence of his profession. There is no basis here for a conversation with philosophers. It is exactly this sort of wisdom that Socrates *dis*claims in the *Apology,* and that disclaimer is the essence of the philosophical profession. For what Socrates proved, if he proved anything in the course of a lifetime of inquiry, is that *no* simple rule is adequate to all situations, that no dogma or obscure revelation or unexamined opinion can with-

---

[11]Jones, *op. cit.,* pp. 299, 301 and footnote pp. 300–01.
[12]*ibid.,* pp. 297, 299.

stand the withering blast of free questioning, and that no belief is worth holding unless it has been submitted for open examination and defended with reasons. (And no life is worth living, he ended by arguing, unless it has met the same tests.) The central claim of Hippocratic medicine, medicine practiced according to the morality of the Hippocratic Oath, is thus the central target of philosophical attack. Philosophy and medicine can cooperate on nothing unless one of those traditions, the Socratic or the Hippocratic, is set aside.

II. SCIENCE AND AUTHORITY

As the claim to wisdom based on secrets handed down from previous generations of practitioners is central to the Hippocratic understanding of medicine, so the accumulated experience of the physician versed in these secrets is the source of all authority over patients and over aspirants to initiation. The orientation of the profession is necessarily conservative; authority rests with the old, the exclusively possessed, the tried and true. In such an arrangement, lay complaints, suggestions for improvement, and all manner of criticism and pressure from patients, other professionals, or the community at large, may safely be ignored or resisted, since no one outside the circle of initiates (unless he or she had illicit access to secret lore) could possibly have anything to contribute to medical practice. As Alasdair MacIntyre reminds us,[13] all such authority has collapsed in the modern era in the West; it is incompatible with the sceptical individualism that dominates the age. The intellectual temper of our time is ruled by Francis Bacon's empiricism, Luther's rebellion, and the radical egalitarianism of the French Revolution; in the disintegration of all tradition-based hierarchy, medical authority has fared no better than political or religious. Hippocratic medicine, at least our own branch of it, died in the nineteenth and early twentieth centuries, a victim of medical science.

The death struggle of Hippocratic medicine in America, as recorded in Donald Konold's *History of American Medical Ethics,*[14] took the form of an intraprofessional conflict between medical practitioners and medical researchers; in the mid-nineteenth century, for the most

---

[13]Alasdair MacIntyre, "Patients as Agents", in S.F. Spicker and H.T. Engelhardt, Jr., (eds.), *Philosophical Medical Ethics, Its Nature and Significance,* Dordrecht, Holland, Reidel, 1977, pp. 197–212.

[14]Donald E. Konold, *A History of American Medical Ethics 1847–1912,* Madison, Wisconsin, State Historical Society of Wisconsin, 1962.

part, "the profession exalted the practitioner over the scientist."[15] In accordance with the operation of such traditions, members of the profession were called upon "to honor the opinions of veteran practitioners above those of doctors newly trained in post-graduate clinics, because experience assured better understandings of medicine than scientific training."[16] Similarly, therapeutic innovations were stubbornly resisted, and the old practices (e.g., bloodletting) were retained quite without regard to empirical studies on their effectiveness. Traditional wisdom, after all, is not supposed to take account of the passing contingencies of individual results.[17] But the climate of the times was progressive, scientific, and at the end impatient with all attempts to halt progress in the name of discipline: the American Medical Association, which sided with the reactionaries as long as it could, was finally forced by dwindling membership to come to terms with the scientifically-oriented sector of the profession.[18] Thereafter it was generally accepted that scientific progress, conducted within the profession or without, had a direct bearing on medical practice, and that it was a positive duty for the physician to keep abreast of that progress and modify personal practice accordingly.[19]

But with this development, the Hippocratic ethic dies. If the fruits of scientific progress, of new discoveries in medical science, are to become part of medical knowledge, the structure crumbles: the body of knowledge is open, not closed; the ancient secrets are not the total, nor the criteria, nor even the heart, of the profession's expertise; experience in practice is less valuable than recently acquired information, and the old practitioners are less valuable than the young, whose information is more recent, and the whole social hierarchy of the profession is overturned. The relevant wisdom is not (only) that given by masters to sworn initiates, transmitted in secrecy, by condescension, from the great to the humble; it is also whatever independent inquiry, by anyone, can turn up. The rewards of professional wisdom go not to the humble, the ones who are willing to obey, conform, and wait in line for the favor of the senior members; instead, they go to the bold, the curious, the ones who are willing to ask questions of anybody and everybody and learn from the answers. The secret society model is discarded in favor of the

---

[15] *ibid.*, p. 33.
[16] *ibid.*, p. 34.
[17] ibid., p. 35.
[18] *ibid.*, pp. 39–42.
[19] *ibid.*, p. 39.

marketplace model; a profession practicing in public, questioning any-one with apparently relevant information, opening up its results for public inspection and comment. The Hippocratic model has been re-placed by one appropriate to Philosophy's founder, the "gadfly" Socra-tes.

To be sure, this new openness obtained originally only among the scientists. Physicians could really only talk to, and learn from, scientists in the health-related fields, many of whom had doctorates in medicine anyway. Now all such scientists could be defined as the "initiates," and all *others* were "the general public." The cause of honesty with the public, especially with patients and their families, was far from gained[20]; as we know, discussion in this area continues to this day.[21] But where concealment continued as common practice, it was for reasons of mere habit or expediency: to hide mistakes, to avoid malpractice litigation, to disguise another physician's mistakes, or simply to present a united medical front for the sake of greater public prestige and influence. The Hippocratic secrecy, secrecy of a religious society as part of the reli-gious obligation of the profession, was gone for good. When concealment ceased to be expedient in the last decades, it was casually dropped or modified; the holy fear had gone out of the guarding.

And now, as is agreeably self-evident, we may openly discuss the problems of bioethics, in an atmosphere that disdains fences and ex-clusivity in professional disciplines. The possibility of a whole new core morality for the profession of medicine arises. But before turning to that possibility, one obvious objection should be disposed of. I have argued that the historical secretiveness of the medical profession is to be derived from a religious orientation, one traceable to the Greeks of the third and fourth centuries before Christ, epitomized in the Hippocratic Oath. I have suggested that this religiously sanctioned exclusiveness continued until it was shattered by the increasingly rapid scientific advance in the nineteenth and early twentieth centuries, and that the move from the Hippocratic toward the Socratic orientation was forced by the change in the nature of medical knowledge, from the traditional

---

[20]*ibid.*, ch. 4.

[21]For example: William D. Kelly and Stanley R. Friesen, "Do Cancer Patients Want to be Told?", *Surgery* 27, 822–826 (1950); E. M. Litin, "Should the Cancer Patient be Told?", *Post-graduate Medicine* 28, 470–75 (1960); Donald Oken, "What to Tell Cancer Patients", *J. Amer. Med. Assn.* 175, 1120–1128 (1961); Samuel Standard and Helmuth Nathan, eds., *Should the Patient Know the Truth?*, New York, Spring, 1955; O. H. Wangestein, "Should Patients be Told They Have Cancer?", *Surgery* 27, 944–947 (1950).

to the empirical. Is it not just as plausible (a cynic might ask) that all this concealment and lying and guarding secrets in "holy purity," among the Greek physicians, was simply the protectionist policy of a group in possession of a saleable commodity that did not want competition? And as for their nineteenth century descendants in the profession, is there any need to call on elements of myth and magic to explain their emphasis on intraprofessional solidarity, collegial silence, and adherence to a self-serving code of ethics? Cannot a simple desire to fix their fees at the highest level, eliminate competition, increase their presence in the legislature, and decrease it in malpractice proceedings explain the same phenomena more concisely? Konold's own evaluation of the history he documents, after all, finds such simple desires perfectly adequate for explanation. But I think such simplicity is misleading. We will always assume, if we have reached the age of common sense and ordinary perception, that every trade, craft, or commercial group will adopt whatever measures seem appropriate to the task of their mutual protection and enrichment, and the medical profession is certainly no exception to the rule. What is distinctive in each case is the choice of policies or strategies toward this end, and not every professional group, even in the learned professions, has attempted secrecy as part of its strategy (witness the academic profession). Moreover, the medical profession's ethic of secrecy, as well as the allied ethic of preservation of the "old" knowledge, continued well past the point when it was even remotely helpful for public relations. The tendency to secretiveness about its own wisdom, in combination with resistance to any change or improvement from outside, was too strong to reduce entirely to the profit motive; it bespeaks an origin in a religious tradition, and the Hippocratic corpus presents us with a likely candidate for such a tradition. Certainly no one would suggest that physicians in the nineteenth century were conscious of the pagan origins of their prejudices, but, as the anthropologists tell us, rite always outlives the myth that justifies it, and the secretive tendency could be transmitted from generation to generation within the profession, by contagion and by example, quite without explanation or justification. But while the tendency was maintained, it should be noted, the revival of the old explanation and ethic, the morality of the Oath, was always possible for the philosophical grounding of the profession's conduct. Now, as we see, that revival is no longer possible. A new ethic, a new core morality for the medical profession, is now in the process of formation. In the remainder of this paper I will attempt a brief outline of a possible form that ethic might take.

III. RESPONSIBILITY IN MEDICINE

We may begin the outline with the suggestion that the medical ethic now developing, the ethic of responsible medicine, will have to satisfy four criteria, none of which could be met by the old morality of the Hippocratic Oath. First, it will have to include in its ambit members of all levels of the profession, those solely employed in research as well as those tending the sick in one-to-one relationship. Second, it must contain essential reference to the expanding body of knowledge that is medical science, and to the desirability of medical progress. Third, taking into account the collective nature of health care at present, it must have as its focus not the moral character of the individual physician, but the goals and activities of the profession as a whole.[22] Fourth, it will have to honor the dignity and autonomy of patients, especially their rights to define for themselves the optimal levels of their own health, to control what happens (insofar as it is under human control at all) in and to their own bodies, and to choose for themselves the extent of their terminal care. These rights of all patients, to choose treatment, to refuse treatment, and to retain privacy and control throughout treatment, are now widely recognized and have commanded a great deal of interest in recent years[23]; since their recognition has an immediate and profound effect on the day-to-day practice of medicine, we shall consider them first.

Following Immanuel Kant,[24] most contemporary philosophers would agree that the essential moral attribute of the human, the precon-

---

[22]Albert R. Jonsen and Andre E. Hellegers, "Conceptual Foundations for an Ethics of Medical Care" in L.R. Tancredi (ed.), *Ethics of Health Care*, Washington, D.C., National Academy of Sciences, 1974, pp. 30–49; and Edmund D. Pellegrino, "Toward an Expanded Medical Ethics: The Hippocratic Ethic Revisited" in R. J. Bulger (ed.), *Hippocrates Revisited*, New York, MedCom Press, 1973, pp. 133–147.

[23]See among many others: George J. Annas, *The Rights of Hospital Patients: The Basic ACLU Guide to a Hospital Patient's Rights*, New York, Avon, 1976; George J. Annas and Joseph M. Healy, Jr., "The Patient Rights Advocate: Redefining the Doctor–Patient Relationship in the Hospital Context", *Vanderbilt Law Rev.* **27**, 243–269 (1974); Robert M. Byrn, "Compulsory Life-Saving Treatment for the Competent Adult", *Fordham Law Rev.* **44**, 1–36 (1975); Norman L. Cantor, "A Patient's Decision to Decline Life-Saving Medical Treatment: Bodily Integrity versus the Preservation of Life", *Rutgers Law Rev.* **26**, 228–264 (1972); Amitai Etzioni, "The Government of Our Body: A Resolution", *Social Policy* (September/October 1973), pp. 46–48; Charles H. Montange, "Informed Consent and the Dying Patient", *Yale Law J.* **83**, 1632–1664 (1974); Judith Thomson, "A Defense of Abortion", *Phil. Public Affairs* **I**, 47–66 (1971).

[24]Immanuel Kant, *Foundations of the Metaphysics of Morals*, Lewis White Beck (trans.) New York, Bobbs-Merrill, 1959; *Kants Werke, Akademie Textausgabe*, Vol. 4, *Grundlegung zur metaphysik der Sitten*, Berlin, Walter de Gruyter, 1938.

dition for all moral obligations, is autonomy—the self-government of
the individual, the free and responsible moral agency of persons. And
if, following Aristotle,[25] we define each thing by its peculiar goodness
or virtue, that autonomy will be an essential element in the notion of
"humanity" itself, the essence of "the person." It is "the person" that
the physician must ultimately treat, not the part of the person's body,
the injury, or the disease. The physician's duty to respect the competent
patient's right to self-government is therefore inherent in medical treatment itself, as well as in civil law and common morality,[26] unless
overidden by some authority higher than any individual autonomy. The
religious traditions, including the Hippocratic, alone could provide
such overiding authority; they are no more, and there is therefore no
conceivable room in medical ethics for a physician's right to thwart or
evade patients rights to govern the courses of their own treatments.
"Paternalistic" models of doctor–patient relationships are simply not
possible at present; that relationship must be governed by a principle
that recognizes the autonomy of both parties. The most reasonable
model for the relationship appears to be the contractual model;[27] the
"contract," by its nature, recognizes the equal dignity, right, and responsibility of the contracting parties, thus honoring the moral assumptions for any free relationship as well as the legal requirements. (The
contractual model can be applied by extension, through advocates and
guardians, in cases of incompetent patients.)

The consequences of the acceptance of the contractual model for
medical care are nothing short of radical: the physician's professional
ethic henceforth contains *no* special obligations regarding the care of
patients. The doctor is bound, in treating a patient, to adhere only to
a common code of decent behavior binding on all men of whatever
profession (the Hippocratic Oath's prohibition on seducing the wife of
a helpless patient would perhaps fall into this category), to observe the
legal conditions of a fair contract (e.g., the prohibitions on the use of
force or fraud, the full disclosure of intent and conditions), and beyond
these general obligations, the content of the agreement on treatment
remains to be worked out by patient and physician in each individual
case. Once a contract has been agreed to, supposing it to fall within
certain very broad legal and ethical bounds (not to create the presump-

---

[25] Aristotle, *Nicomachean Ethics.*

[26] Lisa Newton, "The Healing of the Person", *Connecticut Medicine* **41**, 641–646
(October 1977).

[27] Howard Brody, *Ethical Decisions in Medicine,* Boston, Little Brown, 1976; Robert Veatch, "Models for Ethical Medicine in a Revolutionary Age", *Hastings Center
Report* **2**(3), 5 (1972).

tion of misunderstanding or duress), the physician is morally bound by it, regardless of the custom of the profession or his or her own ordinary procedures in such cases. It is simply not a part of medical ethics, the ethical code that physicians are bound to *qua* physicians, to set the values governing the physician–patient relationship; unless they entail injustice to others, the patient's personal values must govern the decisions on her or his own treatment.[28]

The morality of the developing professional medical ethics cannot govern the physician–patient relationship, which is reduced to a simple contract. Nor can it prescribe the moral character of the physician; the emphasis on the "purity" and "uprightness" of the physician in the old ethic derived, as we have seen, from the ancient belief that knowledge, professional expertise in the field, depended directly on the moral state of the physician's soul. No such belief exists today. It would surely be better for us all if physicians were moral persons; but for no more or different reasons than we would wish all automobile mechanics were moral persons. Medicine is increasingly a public profession: learned in public (at public expense), expanding its knowledge in public discussion, protected from competition (and generally remunerated) by public agencies, and, unavoidably, responsible to the community at large as well as to the private patient.[29] Unlike the small secret society, the public is not primarily interested in the moral character of its members; it is vitally concerned with their actions, especially with the pattern and tendency of those actions over the long run. Nor does our prevalent assumption of the objectivity and public verifiability of knowledge permit any real weight to be given to the physician's personal experience.[30] Hence the professional ethic of medicine moves from an ethic of "virtue" to an ethic of "duty".[31] For the same reasons, it moves from an individual ethic, assuming and binding upon self-employed individuals, working alone with their patients or their research, to a corporate ethic, assuming and binding upon a self-regulating association, setting ethical constraints and value orientations for the profession as a whole, able to give direction to the diversified efforts of individuals in many levels of the profession. As Edmund Pellegrino has argued, the increasingly democratized profession of "health care" (including all health professionals, not just physicians), will have to learn to take corporate respon-

---

[28]Pellegrino, *op. cit.,* p. 137.

[29]Benjamin B. Page, "Who Owns the Professions?", *Hastings Center Report* **5**(5), 7–8 (1975).

[30]George M. Brockway, "The Physician's Appeal to Firsthand Experience", *Hastings Center Report* **6**(2), 9–12 (1976).

[31]Jonsen and Hellegers, *op. cit.*

sibility for fulfilling the purposes that have been assigned to it by the society that supports it.[32]

The major problem confronting the new medical ethic is that of the unity of the profession of medicine. The first effect of the modifications suggested to this point is the disintegration of moral focus: no longer is the morality of the profession contained in the person of the physician, specially (perhaps divinely) qualified, sober, dignified, wise, holding all authority in himself; no longer is it expressed almost exclusively in the dyadic relationship between physician and patient. Instead the morality is outside the physician, set by the community as a whole in concert with the profession as a whole, expressed in an unspecified variety of activities, and the physician is accountable to the community and to the profession for adherence to it. Such a morality can be unified only by subsumption in a single-value system; it can be argued, and the balance of this paper will attempt to argue for the special case of medicine, that every profession in a world that acknowledges no priesthoods must become a single-value enterprise, distinguished from the others by the unique value that defines its activities. Thus the legal profession, at all levels, identifies its activities, jurisdiction, and responsibilities, modifies its professional direction, and informs its education by the pursuit of justice. That is not (certainly not!) to say that every lawyer, judge, and legally trained legislator devotes all her or his time to the pursuit of justice, but simply that the attempt to realize that value is always an appropriate attempt within the profession, that the futherance of justice is always a reason to proceed and the avoidance of injustice always a reason to abstain. Similarly the academic profession pursues truth, the religious pursues sanctity. And by common consensus, the medical profession pursues health, the health of the human person, whether of individual persons (clinical medicine), populations, or future generations. In this value we can find the unity in the activities of medical laboratory research, public health and sanitation, genetic counseling, and personalized care for sick individuals.

Medicine's unifying value is health, the attempt to cure diseases and prevent their occurrence in the future: before proceeding further, three major problems with the assertion must be met. First, what *is* health? A rapidly growing literature attests to the problematic nature of the concept, and suggests that it lacks the coherence to unify anything.[33] Fortunately for us all, the problems do not need to be solved

---

[32]Pellegrino, *op. cit.*

[33]Christopher Boorse, "Health as a Theoretical Concept", (unpublished); and "On the Distinction between Disease and Illness", *Phil. Public Affairs* 5, (49–68) (Fall 1975);

to reach a working understanding of the boundaries of the concept, and such an understanding is entirely sufficient to guide the diversified activities of medicine. We may take Stedman's *Medical Dictionary's* definition of health as "the state of the organism when it functions optimally without evidence of disease or abnormality,"[34] let "disease" and "abnormality" be defined by the AMA *Nomenclature's* listing thereof, and take note that this definition suffices to justify, as "pursuit of health," almost all the activities we would want to identify as "medical." As an operational definition, then, it is quite adequate. To supplement that definition, we may adopt for certain purposes (and to set the maximum range of professional activity) the World Health Organization's definition, "health is a state of complete physical, mental, and social well-being and not merely the absence of disease or infirmity."[35] The WHO definition, however, must be taken with caution and narrowly interpreted, for it can be read to entail the pursuit of *all* human values. Medicine has occasionally been enticed into the pursuit of other values, especially the value of justice (in decisions on the distribution of scarce medical resources, responsibilities to the Third World, etc.), but to no good effect: the coherence of the enterprise depends on a single focus on the multitudinous and complex problems of the world, and compounding the inherent conflicts within medicine by adding conflicts from without, from other potentially incompatible value demands, makes the enterprise impossible.

One persistent conflict within medicine constitutes the second problem with identifying "health" as the ruling value of the medical profession. Not infrequently it turns out that medical knowledge, perhaps immensely helpful for the health of future generations, can best, or only, be acquired by experimentation on living human individuals, to the possible detriment of *their* health. Whose "health" is referred to, in the "pursuit of health?" The answer to this question is quite simple: *everyone's* health is referred to, future people as well as present people, and appeal to the value of health cannot resolve any particular conflicts over the permissibility of human experimentation, research on human subjects. What can be concluded is only that there ought to be guide-

H. Tristram Engelhardt Jr., "Ideology and Etiology", *J. Med. Phil.* **1**, 256–267 (1976); Leon R. Kass, "Regarding the End of Medicine and the Pursuit of Health", *Public Interest* **40**, 11–42 (1975); Stephen R. Kellert, "A Sociocultural Concept of Health and Illness", *J. Med. Phil.* **1**, 222–228 (1976); Joseph Margolis, "The Concept of Disease," *J. Med. Phil.* **1**, 238–255 (1976).

[34]22nd edition.

[35]World Health Organization, *The First Ten Years of the World Health Organization,* Geneva, World Health Organization, 1958, p. 459.

lines for such research: that the guidelines ought to be publicly debated and known and enforced, that they should forbid violations of normal human rights, and that they should favor present people over future people in all cases of irresolvable conflicts. (Even these conclusions, you will note, do not derive from the value of health, but from the value of justice.) The literature on the subject suggests that although such guidelines are difficult to formulate and even more difficult to apply, the task is being addressed by a variety of agencies and is not impossible.[36]

The third problem is a serious one, not to be so easily evaded. It appears in the form of a profound objection to the whole dedication of the profession to the pursuit of a value as abstract and technical as "health." In the abstraction, we may lose the human being, the objectors argue; in the hospital setting especially, but in scientific medicine as a whole, we run the risk of dehumanization in the name of the efficient pursuit of the cure.[37] Essentially, the task of medicine is to treat the person who is ill and seeking help; it is not clear at the outset that the interpretation of the patient's "illness" as a "disease" is always correct,[38] not clear that the physician has a right to demand that the patient allow her or his body to be treated as a depersonalized object, above all not clear that the proliferation of technological miracles and exotic remedies adds one whit to the personal care and individual attention that the patient asks of a personal physician. The objection appears in several forms, more or less far-reaching, ranging from a simple reminder that the patient is a person and needs to be treated as such ("The treatment of a disease may be entirely impersonal; the care of a patient must be completely personal."[39]), to a radical rejection of

---

[36]Mark S. Frankel, "The Development of Policy Guidelines Governing Human Experimentation in the United States: A Case Study of Public Policy-Making for Science and Technology," *Ethics Sci. Med.* **2**, 43–59 (May 1975); U.S. Congress, Senate Subcommittee on Health, *Federal Regulation of Human Experimentation*, 1975, 93rd Congress, first session, May 1975; U. S. Department of Health, Education and Welfare, "Protection of Human Subjects: Fetuses, Pregnant Women, and In Vitro Fertilization", *Federal Register* **40** 40 (Aug. 8, 1975), Part III, pp. 33526–52; U.S. Department of Health, Education and Welfare, National Institutes of Health, "Protection of Human Subjects," *Federal Register* **39** (May 30, 1974); World Medical Association, Declaration of Helsinki, "Recommendations Guiding Medical Doctors in Biomedical Research Involving Human Subjects", *World Medical Journal,* (September, 1974).

[37]Pellegrino, *op. cit.,* p. 139.

[38]Eric J. Cassell, "The Organ's Disease, the Man's Illness and the Healer's Art", *Hastings Center Report,* **6**(2), 27–37 (April 1976).

[39]See Francis Weld Peabody, "The Care of the Patient", *J. Amer. Med. Assn.* **88**, 877 (1927), reprinted in R. J. Bulger (ed.), *Hippocrates Revisited,* New York, MedCom Press, 1973, pp. 50–60 (see especially p. 51).

the entirety of medical science and the medical model itself as fatally destructive of human capacities to cope with the human condition.[40] In the form of the reminder, the objection is entirely acceptable. Surely, as above, the human person as a whole is the ultimate beneficiary of medical care; but none of this is incompatible with a unique professional commitment to the value of health. Real conflict arises only with the form that seeks to redirect the efforts of the practitioners away from the cure of disease to something else: "Medicine is the science of care, not the science of cure," as one physician[41] put the claim and from a fundamental commitment, not to health, but to personal care of the patient, a set of ethical constraints and directives can be derived that are totally different from those derivable from a commitment to health: for instance, in its constraints, it rules out research on human subjects, at least in a medical setting. It discards the first two criteria suggested at the start of the section: it applies only to practicing physicians, since those engaged in scientific research never deal one-to-one with patients (except as experimental subjects), and it makes no reference to the body of knowledge the physician has at his disposal. In its directives, it is holistic and humanistic, requiring of the physician a mature compassion, sensitivity to the deep and often inarticulate needs of the patient, a concern for the patient in all aspects of his or her personal life and work, and a willingness to spend the time and effort in the one-to-one relationship of the medical encounter to help the patient achieve a personal growth and re-integration beyond the mere cure of the disease. As Richard Magraw puts it,

> "I think we will not be much off the mark if we say that the best generalization we can make about personal health service is that the doctor's fundamental task is to help the people who came to him in any way he can. He is to alleviate the pains of living or lighten in any way he can the anguish of existence."[42]

This objection requires serious consideration for its insight into the discouraging lack of real personal respect and close personal relationships elsewhere in contemporary society, its recognition that the need for these may be at the heart of the malaise, the "illness," of each patient, as it is of our times. There is no doubt of the worth of its moral

---

[40]Ivan Illich, *Medical Nemesis: The Expropriation of Health,* New York, Random House, 1976.

[41]Robert E. Cooke, "Whose Suffering?", *J. Pediatrics* **80,** 906–908 (1972).

[42]Richard M. Magraw, "Science and Humanism: Medicine and Existential Anguish," in R. J. Bulger (ed.), *Hippocrates Revisited,* New York, MedCom Press, 1973, pp. 43–49.

ideal. The question, however, is not whether it is worthwhile, nor whether "personal care" is a proper defining value for any profession; the question is whether it is superior to "health" as a defining value for the medical profession. The answer here seems to be No: for a variety of reasons, one theoretical and the rest proceeding from practice.

From the practical standpoint, it appears that physicians as currently prepared are not qualified or inclined to devote their professional lives to personal care; that nurses as currently prepared *are* so qualified and inclined, hence a reorientation of the medical profession is not necessary; and that a commitment to "care" alone in an area of rapidly changing knowledge, ignoring the obligation of competence in applying that knowledge, can end up doing the patient more harm than good. Taking these one at a time: most obviously, the practice of medicine along the holistic and humanistic lines suggested would require very sensitive physicians with a good deal of time to spend with each patient. There are not enough physicians for this type of practice, and the ones we have, with some exceptions, possess neither the ability nor inclination to practice medicine in this way. This reply is of course not sufficient in itself: the features of the profession mentioned could certainly be changed, and would be, by modifying medical education, if we decided that humanistic medicine were our first priority in the profession. The second reason for rejecting that alternative is that the redirection of physicians and medical education may not be necessary to restore a measure of humanism to patient care; a more fruitful course of action might be to turn more of the decision-making authority for patient care over to the nurses, since the nursing profession already acknowledges personal care as its primary commitment. Nursing seems to be the "science of care" the humanists were looking for. The third practical reason to reject the humanist's plea for a profession focused on caring, not curing, is that such a focus ignores, and ignores dangerously, the potential for harm in ignorance, however well-meaning; there is no necessary reference, in the humanist's definition, to an obligation to keep up with the progress in medical knowledge, the obligation of competence. As Edmund Pellegrino puts it,

> The patient has a right to access to the vast stores of new knowledge useful to medicine. Failure of the physician to make this reservoir available and accessible is a moral failure . . . competence has become the first ethical precept for the modern physician after integrity. It is also the prime humane precept and the one most peculiar to the physician's function in society.[43]

It simply is not humane, or moral, to do damage when knowledge is available to do good. But that last assertion, that competence is "peculiar to the physician's function in society," calls our attention to the theoretical objection: the conceptual linkage between medical practice and the science that grounds it. What counts as *medical* care, as opposed to other kinds of care, is distinguished by access to that body of knowledge known as medical science; the physician's care specifically applies that science or fails altogether as care. The conceptual linkage is unavoidable once the bond with the Hippocratic tradition is broken; if the physician's qualification as healer (i.e., identity as a physician) cannot rest on sanctity or personal experience, it *must* rest on mastery of that body of science, i.e., of that science in its most current form, and the ability to apply it in an individual case.

Even in theory, then, the non-Hippocratic physician cannot be separated from medical research. The professional ethic therefore satisfies the first two criteria for responsible medicine. "The profession" is to include *all* those involved in the enhancement of health, in their relevant activities only: from scientists engaged in research on the cause of diseases to paramedics engaged in rescue work. And the relevance of the activity will be determined by its contribution to or reliance upon a body of knowledge, the limits of which will be determined by the criterion of usefulness in the restoration or preservation of someone's health, now or in the future. As Toulmin has pointed out, this body of knowledge varies in type from the most abstract and theoretical to the most particular and personally committed.[44] At the level of medical science, it partakes of the precision and certainty of "pure" science; at the level of personal knowledge, the physician's knowledge of the patient, it is necessarily inexact and fallible, as is every science of particulars.[45] Throughout the levels, the only unity is that imparted by the common value: at the physician–patient level, the physician's commitment to the patient's recovery is the best guarantee the patient has in an essentially uncertain endeavor;[46] at the research level, the commitment to the eradication of disability and disease distinguishes medical research from other types, and directs the scientist's choice of projects.

A professional ethic for responsible medicine is then an ethic of value-oriented knowledge at multiple levels; a final, and crucial, feature of this knowledge is that each level is utterly dependent on free communication with the others for its own proper functioning. The major directive of the ethic is quasi-educational in nature, requiring the con-

[43]Pellegrino, *op. cit.*, p. 140.

[44]Stephen Toulmin, "On the Nature of the Physician's Understanding", *J. Med. Phil.* **1**, 32ff (1976).

[45]*ibid.*, and Samuel Gorovitz and Alasdair MacIntyre, "Toward a Theory of Medical Fallibility", *J. Med. Phil.* **1**, 51–71 (1976).

[46]Gorovitz and MacIntyre, *ibid.*, p. 64.

stant transmission of value-oriented information from laboratory to clinic and back again. The information is scientific: a research team announces that a certain chemical appears to halt the action of viruses, a clinic informs the research team that the chemical is, or is not, effective in the cases they have seen. The value-orientation is the common purpose that obligates scientist and clinician to the communication, and determines the activities that are appropriate in the light of the information. The centrality of this medical dialog, this value-oriented transmission of information, calls for a special place in the new medical ethic. We can say that this dialog is the locus of the moral obligation of the profession, replacing the person of the physician. The maintenance of the dialog, its content of truth direction toward usefulness, is the first duty of the medical profession.

An essential element in this dialog is the inclusion of patients, who for the purposes of their own treatment only, thus become members of the medical profession. If the claim seems paradoxical at best (since the patients originally arrived at the physician's door precisely because they lacked the expertise that defines the profession), consider that, during the course of the treatment, the patients must be (by the requirements of the criterion of autonomy) fully responsible participants in all identifiably medical activities that are undertaken in their cases. The patients must understand, and agree to, the use of any drugs, or any medical or surgical procedures involved; obtaining their agreements (under the conditions of "informed consent") will require them to understand exactly as much anatomy, physiology, and chemistry as is needed for a responsible evaluation of the consequences foreseen from the treatment. To satisfy the requirement of autonomy, the physician must educate the patients in the art and science of medicine to the point that they can make responsible choices; this point alone is a measure of how far we have moved away from the Hippocratic ethic of secrecy. As a member *pro tem* of the profession, the patients too are obligated by its ethical requirement of full and free communication, as well as the contractual requirement of faithful performance of what has been agreed upon (e.g., to follow a regimen, administer drugs), as part of the commitment to the restoration of health (his or her own). The patients are the necessary start and finish of the pattern of exchange: it is their pain and urgent need that constantly renews the whole profession's commitment to the eradication of its cause; it is their testimony that ultimately decides the usefulness of the medical scientist's discoveries and the physician's techniques. What distinguishes patients from other health professionals is not any qualitative difference in their knowledge and actions during their own treatments, but the temporary nature of

their involvement; as soon as the contractual/educational bond is sev-
ered by reason of recovery or other indication for termination, that
participation ceases. The patients are therefore never under the primary
obligation of the physician: to keep abreast of the newest knowledge in
the field, to renew personal competence by constant communication
with the other levels of the profession. A patient's education and commitment to the value of health must be initiated again in each new
medical encounter.

The final criterion suggested for a workable professional ethic of
medicine was that the ethic must be corporate, containing a notion of
responsibility applicable to the profession as a whole. I am unprepared
at present to suggest institutional means and measures for effecting that
requirement in operation. But I believe a framework for a notion of
corporate agency can be realized in practice from the experience of a
profession carrying out the directive elaborated above. When the first
directive of a profession is that of multi-level communication, the profession is a corporate enterprise to begin with. Procedures set up in the
course of the enterprise to facilitate that communication could perhaps,
according to time and circumstance, be modified to monitor the application and effect of the information communicated. A monitoring procedure could perhaps, according to time and circumstance, be modified
to set up an agency charged with overseeing the enterprise, authorized
to modify it (and discipline its members when necessary), responsible
for its ultimate faithfulness to the value that defines and directs it. All
of this is supposition: the moral and conceptual groundwork is in place,
the legal and social superstructure will no doubt take time to construct.

IV. CONCLUSION

The professional ethic of the Hippocratic tradition is dead; long live the
professional ethic of responsible medicine. An ethic contained in the
morally virtuous person of the physician, qualified as a healer by the
possession of secret wisdom acquired intact from previous generations
and enriched by their experience, applied to ignorant and trusting
patients, is gone. It is supplanted by an ethic now in formation, contained in the medical dialog of the profession as a whole, qualified for
healing by their open wisdom, acquired from continual inquiry and
never complete, openly shared with responsible patients who join with
them in the enterprise of medicine, an enterprise working toward a
corporate framework for the exercise of its public responsibilities. Participation in, and enhancement of, that dialog is the clearest obligation

of the physician, fulfillment of which qualifies her or him to contract with patients for the performance of such medical procedures as may be agreed upon. The autonomy of the patient, the dedication of the scientist, and the ability of the physician to teach and learn from both of them, are the elements of responsibility in medicine.

Comments on "Hippocrates Lost, A Professional Ethic Regained: Reflections on the Death of the Hippocratic Tradition"

Donald E. Zarfas

Department of Psychiatry University of Western Ontario, and Children's Psychiatric Research Institute, London, Ontario, Canada

Dr. Newton begins her paper with the concept that the Hippocratic tradition is no longer with us. I am not sure that she is totally justified in making this statement. She sees the medical profession in the past as having been a largely secret religious society, closed to outsiders, deceitful to its patients, and nonscientific in nature. In some ways, she is correct. She is correct in saying that this has changed, and I am sure that there are few, if any, of us who would like to perpetuate those aspects of the old tradition. That is not, however, to deny that the tradition of secrecy in the past may well have served the physician of the prescientific era. Of the charisma and the mystique, which may well have been the doctor's most therapeutic tools, some aspects still play a significant role in therapy. People feel better simply by having made a contact with a physician, or even with a clinic. As a matter of fact, some of those who are on the waiting list to be seen in psychiatric clinics do extremely well. Perhaps this is because they have now begun to recognize their problems and to work at them themselves. In some of the less sophisticated communities in developing countries, one is still able to appraise the role of witch doctors and the magical treatments believed in by their "patients." Sometimes, these are reported to be more effective than are certain of our modern medicines, so that even this aspect of the Hippocratic tradition may have served a less sophisticated society advantageously.

Many universities administer to the graduating class an oath referred to as the "Hippocratic Oath," but one, of course, that is much changed from the original. Other schools require their students to adhere to a different code. Their aim is to provide the new physician with guidelines that will ensure a high quality of medical care for the patient, and, as the codes indicate, maintain the honor and the noble tradition of the profession. The International Code of Medical Ethics, and the Declaration of Geneva (which was approved by the World Medical Association and adopted by member organizations), go a long way to modernize the ethics of the profession in the direction of the changing moral attitudes of society. These codes and the Health Disciplines Act of the Province of Ontario address most of the important moral issues and they guide the physician in his or her behavior and responsibilities toward the patient, setting out clearly those practices that constitute professional misconduct. The lawyers Morris C. Schumiatcher and D. Jur write: "the physician has been trusted over the years because of the ethical and moral standards he subscribed to in taking the Hippocratic Oath . . . the patient knows he is neither going to be ripped off or ripped up by his doctor. He also knows that the doctor will take all due care and all reasonable measures to make him well and not do him in."[1]

But society and its mores change, and it is the responsibility of medicine to respond to these changing attitudes. A criticism long leveled at medicine has been that it changes slowly. People say that it is too traditional, too protective of the past, too uncritical of its members, yet, in the eyes of many others, including at least some professionals in medicine, it has moved much too quickly and far in recent years. Surely the issue concerning abortion is one example in which the question of right or wrong has not been fully answered for many physicians, despite the fact that legal prohibitions to this practice have been removed in some jurisdictions. Similarly, the question recently posed by the Committee of the Anglican Church regarding euthanasia of seriously-retarded children is surely one of terrifying import for philosophers, lawyers, and physicians. It may be at these times that we need the old Oath more than ever, since it spoke with such certainty

---

[1]M. C. Schumiatcher and D. Jur, "Medical Heroics and the Good Death," *J. Can. Med. Assn.* **117**, 520–522 (1977).

about the prohibition of abortion and the prescription of deadly medicines. The new Oaths, although not silent on these issues, seem to open the way to a much more ambivalent interpretation, lending arguments to both supporters and opposers of legal abortion.[2]

With regard to Dr. Newton's new ethic of responsible medicine, we may say that certainly all levels of the profession, from the basic scientific researcher to all varieties of clinician, are utilizing the rapidly expanding knowledge brought to it through the medical sciences. Similarly, there seems to be little doubt that health must, as she indicates, constitute the unifying value of medicine. As a psychiatrist, I must encourage the profession to be increasingly concerned about the mental and social well-being of its patients and of society. Therefore, I would tend to subscribe somewhat more to the broader interpretations of health encouraged by the World Health Organization, in spite of recognizing the virtual impossibility of ever achieving universal and complete physical, mental, and social well-being in our time. Such concepts, however, are important in that they alert us to the breadth of such challenges as the prevention of disability: for example, the poverty found in the core areas of North American cities, and in wilderness areas among some of our native and disadvantaged people, make an immense etiological contribution both to mild mental retardation and to borderline intelligence with the life-long disability they bring. The underutilization of existing Health Services by people in both of these kinds of environment contribute to poor physical health among their populations: some social milieus contribute immensely to alcohol and drug abuse and, ultimately, to crime. The health team and the social service agencies must address themselves to this broader interpretation of "health."

Moving from this global concept of health to its application to the individual, Dr. Newton quotes Dr. Robert Cook, one of America's most famous paediatricians, as saying "Medicine is the science of care, not cure." I, personally, would have been more comfortable with Dr. Cook's remarks if he had said "Medicine is the science not only of cure but also of care." He meant, I believe, that therapy without cure is extremely worthwhile if the therapy contributes to the welfare of the individual. As Richard McCraw (also quoted by Dr. Newton) stated: "The doctor's fundamental task is to help the people who come to him in any way he can." For me, this justifies WHO's definition of health, since it allows us to see it from the glo-

---

[2]Compare Loretta Vicente, "Declaration of Geneva: Some Implications," *Can. Doctor,* September 1977.

bal, as well as from the individual's, viewpoint. The crux of medicine is the physician's responsibility to the patient here and now; and all of the codes support this view. The patient's treatment must be personal and comprehensive: the medical professional must adopt a holistic approach. This, I believe, bridges the gap between personal care and health seen by Dr. Newton. But in so saying, we must admit that medical schools do provide more training for cure than for care. However, this is now being recognized, and courses today are beginning to reflect this recognition.

Dr. Newton has indicated that perhaps more of the decision-making authority for patient care should be turned over to nurses, since nurses seem to be accepting the science of care as their goal. I am uncertain whether or not she sees them playing the role of advocate for the patient, helping the individual to understand and decide when, what, and how much treatment she or he needs. I wonder whether many nurses really are prepared to assume this role and carry it out without bias.

Lastly, Dr. Newton has indicated the importance of communication between doctor and patient, and the involvement of the patient in the decision-making process. She has also introduced the concept of a corporate body of medicine to control and direct its members in the provision of individual medical care. Since these ideas are, in my opinion, closely interwoven, I will attempt to comment on both of them at the same time. First of all, physicians are increasingly informing patients of the natures of their disorders, and of the merits and risks of the various treatment-plans open to them. The whole concept of informed consent, which has become an integral part of medicine in the United States, and is becoming so in Canada, requires this communication. I see it as being a simple process in some clear-cut circumstances. But I wonder whether it is always as simple a process in the more complicated conditions in which diagnosis, treatment, or management is less well understood and less well defined. Contractual arrangements between physicians and patients are, of course, becoming popular in medicine, even in psychiatry and other areas in which indications are seldom clear-cut. But contracts can be negotiated without full and open communication, and I believe there are still many circumstances in which physicians do, in fact, withhold information from the patient simply because they feel that full knowledge would not be in the best interests of the patient. For example, suppose that in a genetic screening of a fetus or of a newborn infant, the XYY chromosome abnormality, once called the "crime chromosome," is found. This information would appear clearly to be information to which the parent has the right of access. Such information, however, may have a distinct influence on the way the parents interpret the behavior of their developing child and, therefore, perhaps, influence their rearing practices to the detriment of the child. In such a case the giving of information may do more harm than good. Thus, the ethical implications for or against full communication are not always obvious. Recently, and in a similar vein, people are beginning to ask who owns the patient's medical record. If patients are granted the full right to view the information in their charts, then I believe that the information that will appear in such charts may only be that which the physicians and health teams feel is suitable. A comparable situation exists in education, since in some jurisdictions the student record must be made available to the parents. In such circumstances, do we in fact have two or perhaps more records? We have those that appear in the Official Records. We also have those that appear in educators' or doctors' heads and in their private files. This practice may, in fact, be worse both for the student and for the patient in the long run.

Corporate medicine is surely with us today in Canada. To visualize an even more highly structured Health Care System, short of one that is completely state-owned and operated, is indeed difficult. There are few, if any, doctors who can avoid involvement with the authorities, both within the profession and in the government. Many of those who are leaving Ontario to practice in the United States hope to find less interference in the manner in which they deal with their patients: this is probably wishful thinking. Today corporate medicine determines our fees and the numbers of patients that a physician can see. Government computers record for each patient's visit the diagnosis and the methods of treatment; so what happens in the consulting room is no longer a secret between the physician and the patient. Goodness only knows how many other people now have at least some entry into what was previously entirely private information. In Ontario, I believe that neither the doctor nor the patient have any need today to worry about confidentiality. But what about the future? In Britain, which has a more socialized and controlled health system, physicians are told where to practice, and patients are told which doctor to see. In Russia, where government control reaches the ultimate, we hear that psychiatric hospitals are being used for treatment of political dissidents. I am sure this latter is not what Dr. Newton visualized as "Corporate Medicine." I subscribe with most other physicians to organized and responsible medicine. But I am concerned that in removing individual autonomy from the physician, one makes it so much easier for Government to accept full control over private practice and, indeed, over the entire operation of the health care system. In a less democratic country, we have seen to what extent this can lead. Let us hope and strive to be sure that such institutional means are not employed as could spell an end to what we have grown to accept and respect in the Hippocratic tradition.

Physicians as Body Mechanics

Michael D. Bayles

Department of Philosophy, University of Kentucky, Lexington, Kentucky

Much has been written about the physician–patient relationship. The traditional relationship, many people claim, has broken down under the conditions of modern medical practice, and it is not yet clear what will replace it. Many writers on medical ethics consider the relation to be a special one, not paralleled elsewhere in society. This paper aims to downgrade that special claim without denigrating the profession of medicine. The occupation of auto mechanic has arisen in society almost simultaneously with the progress of medicine and the breakdown of the traditional physician–patient relationship. This paper compares the automobile mechanic–owner relationship to the physician-patient relationship. Despite one's initial aversion to this analogy, it soon seems a very strong and informative one.

However, a preliminary caution is needed. There is no single appropriate model of the physician–patient relation. There are at least several important variations, depending on the context. Nevertheless, the impulse of philosophy, undoubtedly given too free a rein in the past, is to generalize. The focus here then will be upon one common medical context and physician–patient relationship, that of a patient with a nonfatal physical disease seeing a primary care physician. Particularly excluded are the relations of patients to psychiatrists and to health care teams.

I. WHY SEE A MECHANIC?

There are three basic reasons why reasonable persons may have others make judgments on their behalves. First, although capable of doing so,

167

one may not wish to be bothered making them. One will allow another to do so if that person will make them as oneself would, or if the differences are insignificant. For example, one authorizes a secretary to order any needed office supplies. Second, the judgments may require knowledge or expertise one does not have. For example, a tax accountant knows the tax laws. It is not necessarily the case that one is incapable of acquiring the needed knowledge, but it is not worth the effort. No one can acquire all the knowledge needed for all the activities and decisions of modern life. One primarily goes to an expert for information and service. Third, one may allow others to make judgments if one will be or is already mentally incompetent. Thus, some people voluntarily enter mental hospitals.

The ordinary person visits an auto mechanic or physician for the second reason: lack of personal knowledge about how to diagnose or repair a misfiring engine, or how to treat fever and nausea. At least four basic models exist of the relation between nonexpert and expert. *(1)* In the agency model, an expert is a technician who acts in behalf, and at the direction, of a nonexpert. *(2)* In the model of a contract between equals dealing at arm's length, an expert sells services just as a grocery store sells food. *(3)* In the paternalistic model, an expert with superior knowledge and skills stands to the nonexpert as a parent to a child. *(4)* In the fiduciary model, although there is a contract, it is not between equals at arm's length. Because of the nonexpert's inferior knowledge and vulnerability, an expert has a special obligation to consider the nonexpert's interests.

Which model is appropriate depends upon the characteristics of the people involved and the situation in which the relationship develops. The relationship between automobile mechanic and auto owner is obviously not that of paternalism. It would be gratuitous for mechanics to treat as children adult owners with the mental ability and experience to fend for themselves in society. Nonetheless, the paternalistic model has been and is still applied to the physician–patient relationship. However, in recent years it has been widely criticized and rejected as inappropriate for most situations. Rather than refight that battle, the remainder of this paper focuses on the other three models.

Several authors have provided reasons for rejecting the agency and contract models for the physician–patient relationship.[1] Agency as-

---

[1]See, for examples, Roger D. Masters, "Is Contract an Adequate Basis for Medical Ethics," *Hastings Center Report* **5**, 25 (December, 1975); William F. May, "Code, Covenant, Contract, or Philanthropy," *Hastings Center Report* **5**, 35 (December 1975); and H. Tristram Engelhardt, Jr., "Rights and Responsibilities of Patients and Physi-

sumes that the principal knows what needs to be done and directs the technician or agent in doing it. The principal is in a superior position to the agent. Contract generally assumes bargaining between equals. At least four factors are frequently cited for rejecting these assumptions in the physician–patient relation. *(1)* The physicians' knowledge far exceeds that of their patients. *(2)* The patients are concerned and worried about their health and have more at stake than physicians, who can find other patients more easily than the patients can locate alternate physicians. *(3)* The patients cannot shop around for another physician; the market mechanism does not work.[2] *(4)* Third party payers (insurance companies) limit the services, or amounts, for which they will pay.

However, these same four factors also generally apply to the automobile mechanic–owner relationship, thereby rendering the agency and contract models equally inappropriate for it. Consequently, to the extent that they support applying the fiduciary model to the patient–physician relationship, they also support applying it to the mechanic–owner relationship. *(1)* The mechanic and owner are not equally knowledgeable. The mechanic possesses technical information and expertise that many owners lack. Many owners do not know what a distributor cap is, let alone the functioning of an alternator or the automatic transmission. Evidence for the general ignorance of owners about automobiles may be found in the number of unneeded shock absorbers and tires sold to tourists along old Route 66 in the Southwest. *(2)* An owner may be psychologically disturbed by anxiety about an illness, and have more at stake than the mechanic. When a $5000 automobile will not start or run properly, the owner is psychologically disadvantaged vis-à-vis the mechanic. The owner needs the car to get to work and complete essential shopping. The mechanic does not have a comparable or urgent need. An owner has more at stake financially than the mechanic. It costs an owner more to have a car repaired than a mechanic makes in repairing it. *(3)* An owner is often limited in seeking repairs elsewhere. There may be no opportunity to shop around for the best buy, as the contract model assumes, and indeed the owner may need the car repaired as soon as possible. If the car will not run at all, then to go to another mechanic involves having the car towed. Moreover, often there are only one or two mechanics competent to

---

cians," in *Medical Treatment of the Dying: Moral Issues* M. D. Bayles and D. M. High, eds., Cambridge, Mass., Schenkman, and G. K. Hall 1978, pp. 9–28. See also Cobbs vs Grant, 8 Cal. 3d 229, 502 P.2d 1, 9, 104 Cal. Rptr. 505 (1972).

[2]See also, Kenneth Arrow, "Uncertainty and the Welfare Economics of Medical Care," *Amer. Econ. Rev.* **53**, 948–954 (1963).

work on a particular type of car. Many mechanics may lack the necessary training, tools, or parts to do the work. *(4)* Third party payers are often involved, and these of course restrict the bargaining freedom of both parties. Although automobile insurance companies frequently allow the insured to obtain estimates for accident repairs, they may pay only the lowest amount.

Charles Fried argues against the analogy between the physician–patient and the mechanic–owner relationships.[3] His chief contention is that the patients' relationships to their bodies are different from the owners' relationships to their automobiles. The latter relationship is one of property, but not the former. People's bodies are not productive goods they own, but part of the persons themselves. Although Fried's claim about the difference between one's relationship to one's body and automobile is correct, it does not necessarily follow that there is or should be a difference between one's relationships to physicians and mechanics. The range of interests that the physician's treatment may affect is often much broader than that of those affected by a mechanic. Moreover, what a physician does may more directly and immediately affect one's mental state (e.g., the administration of painful treatments or drugs affecting one's mental state). These factors affect the degree of concern a physician should show and the manner of treatment, rather than the kind of relationship between the expert and nonexpert. Moreover, Fried contends that the physician–patient relationship should be viewed as a fiduciary one,[4] and the argument here is that the mechanic–owner relationship should also be viewed as a fiduciary one.

This objection, like most of those that follow, is based upon alleged differences in the relationships of the experts or nonexperts to the subject matters of their transactions. However, this form of objection is mistaken in principle. Differences in the relationships of experts (E s) or nonexperts (Ns) to the subject matter of their transactions or relationships does not imply that the relationships between them vary. If E and N have a relationship with respect to S, it is the differences between E and N, rather than differences in their separate relationships to S, that are primary for the nature of the relationship between E and N. However, since these objections also generally exaggerate the differences between the relationships of owners to automobiles and patients to bodies, or of mechanics to automobiles and physicians to bodies, their factual grounds should be briefly considered.

---

[3] *Medical Experimentation,* Clinical Studies, Vol. 5, New York: American Elsevier Publishing Co., 1974, p. 96.

[4] *Ibid.,* p. 34.

A second ground for asserting disanalogy is the comparable importance of the interests of patients and owners. A patient's interest in health, maybe even life, is at stake, whereas an owner chiefly has a financial interest involved. This difference is often not as major as one might expect. On the one hand, for many people an automobile is not a mere luxury; it is an essential element of transportation necessary for employment, getting food, going to the doctor, and so on. Also one might risk one's life in the hands of a mechanic. Faulty repairs or failures to detect malfunctioning may lead to accidents causing injury or death. On the other hand, not every illness for which people see physicians is life-threatening. Writers on medical ethics often focus on dramatic, fatal illnesses, which is a mistake comparable to that of philosophers of law focusing upon rare Supreme Court cases. Most medical problems, like most legal cases, are less important—the flu, a sprain, a bronchial infection, a hernia, a broken arm, and so on.

Another objection to the analogy might be based upon an alleged difference between medical and mechanical science. Some philosophers claim that the concepts of health and disease, and consequently medical science, are value-laden, whereas values may not be crucial to determining the proper functioning of an automobile.[5] One cannot define health or disease without reference to the values of the patient or society. Consequently, although a physician must necessarily make value judgements, a mechanic characteristically does not. This argument over the nature of physicians' activity is not that important. Mechanics must also consider owners' goals and values in determining whether an automobile functions properly. For example, one owner may desire a very smooth and comfortable ride and want shocks replaced, although another owner would be quite happy with the rougher ride. Social values may also be relevant, as when there is emphasis upon the most efficient gas consumption to conserve energy. Moreover, it is not clear that the concepts of health and disease are indeed value-dependent. Finally, at best the argument shows that the way in which values are related to the judgments and activities of physicians differs from the way in which they are related to those of mechanics, but that fact only increases the importance of physicians considering patients' values and of their obligation to do so.

A final objection to the analogy is that although one often goes to

---

[5]See both Joseph Margolis, "The Concept of Disease," and H. Tristram Engelhardt, Jr., "Ideology and Etiology," in *J. Med. Phil.* **1**, 238–268 (1976). Cf. Christopher Boorse, "On the Distinction between Disease and Illness," *Phil. Public Affairs* **5**, 49–68 (1975).

a mechanic for maintenance, one rarely goes to a physician for maintenance. Cars are not always in need of repairs, but they need oil changes,
lubrication, and tune-ups. However, people are not always ill when they
go to a physician. They may merely want an eye examination or Pap
test. Preventive medicine is concerned to provide just such "scheduled
maintenance." If people go to physicians for preventive medicine less
than they do to mechanics for preventive maintenance, it may be because they are more concerned and worried about their cars than
themselves. Also, tune-ups and other such services are not mere preventive maintenance; often the engine is not working at full efficiency when
they are performed. A change of oil or antifreeze might be comparable
to taking vitamins.

II. WHY TRUST A MECHANIC?

If an analogy between the mechanic–owner and physician–patient relationships is thus plausible, one might get a handle on the analysis of
physician–patient relationships by considering what an automobile
owner desires or expects of a mechanic. Probably the most asked question is: can this mechanic be trusted? One wants a mechanic who is
trustworthy to look after one's interests in the automobile, and one
wants a mechanic one can trust to protect and further those interests.
Thus, the virtues of a mechanic are those that comprise trustworthi-
ness. And since trust is crucial to the relationship, it is best described
by the fiduciary model.

Physicians emphasize the concept of patient trust. However, when
some physicians speak of patients having trust, they mean trust in
physicians to make decisions about and to provide appropriate treatment. In short, they mean patients should trust physicians as children
should trust parents. The element of trust involved in the fiduciary
model is different from that of the paternalistic one. One must always
ask, "Trust to do what?" A virtue of the analogy between the mechanic–owner and the physician–patient relationships is that it provides a
new perspective on what patients might trust physicians to do.

Although many of the ethical or moral features of the physician-
patient relationship derive from a physician being worthy of a patient's
trust, it is a mistake to think the whole of medical ethics can be derived
from this source. The physician–patient relationship can at best ground
the ethical obligations between these two parties. Physicians, like automobile mechanics, have obligations to others that do not arise from the
relationship. These obligations depend on the role of the profession or

occupation in society. Also, there will at least be variations in the fiduciary relationship, depending on the situation; and in some situations (e.g., that of an unconscious emergency patient) another model, such as the paternalistic one, may be more appropriate. Only a few ethical elements of the physician–patient relationship can briefly be considered here by use of the automobile mechanic–owner analogy.

In New York state, automobile repair shops, like physicians, are licensed. The explicit purpose of such licensing is to indicate that the shop and its mechanics are worthy of trust. (Of course, implicitly, licensing is frequently advocated by professions and occupations in order to restrict admission and increase fees or prices.) One major point of licensing is of course to certify competence. Mechanics must know what they are doing or they may make matters worse. Incompetence may be grounds for suspending their licenses.

Two interesting points arise from the licensing of both mechanics and physicians. First, no one seriously argues that mechanics should be a self-regulating group. Public control by the state is readily assumed to be appropriate. However, although there is some control of physicians by the state, there is much emphasis upon self-regulation by the medical profession. The analogy suggests that self-regulation has at least been overemphasized. Second, repair garages may be fined or have their licenses suspended or revoked for overcharging automobile owners. There is no parallel practice with physicians; overcharging patients is not a widely recognized reason for suspending or revoking their licenses. By analogy, however, physician "theft" in the form of overcharging or performing unnecessary services should be a ground for so doing, and New York is currently considering such a change in the law.

The mechanic–owner analogy also provides a basis for analyzing two much discussed issues in medical ethics, namely, a physician's obligation of full, truthful disclosure and patient consent. When an automobile owner or patient goes to a mechanic or physician, that person is seeking the latter's expert knowledge. An owner wants a mechanic to describe what is wrong with the car and then, usually, to repair it. A person goes to a physician to obtain a diagnosis of an ailment and a recommendation of what drug to take, surgery to have, or other treatment regimen to pursue. To be worthy of trust, the mechanic must make a full, truthful disclosure of what has been found. Similarly, to be worthy of trust, a physician must make full, prompt, and truthful disclosure.

A physician's obligation of full, truthful disclosure is not absolute. There may be, or at least many physicians believe there may be, occasions on which it is outweighed by other obligations. The most frequently discussed situation concerns informing a patient of terminal illness, of cancer in particular. The argument for nondisclosure is that the information would be harmful to the patient's well-being, particularly the individual's mental well-being.[6] On what is this obligation to the patient's well-being based? Physicians usually rest it upon the traditional principle that a physician's first obligation is to do no harm. This principle merely moves the issue back one step. The problem then becomes whether the information and psychological distress are, in the total situation, harm. Although the patient may suffer psychological depression, he or she may also want to know the truth.[7] One might compare the situation to that confronting an automobile mechanic who discovers that an automobile needs such major repairs that it would be cheaper to buy a new one. Suppose the mechanic does not tell the owner, because it would produce psychological distress and nothing can be done to fix the car anyway. The mechanic might even contend that the owner either "really knows" the car is shot, or else does not want to know. In that case, the mechanic has clearly stepped beyond the bounds of the relationship and failed in being responsible to the owner.

These comments are not intended to suggest that a physician is never justified in withholding information from a patient. A physician may certainly refrain from giving a full explanation to a patient immediately after a heart attack, only to give it after the patient has recovered sufficiently. Similarly, a physician can often harmlessly tell a cancer patient of the nature of the disease by waiting for a time when the patient is psychologically ready to grapple with the problem. Never providing a patient important information is extremely difficult to justify; it is easier to justify at least some delay before providing the obligatory information.

The trust one may repose in a mechanic is to provide accurate, relevant information and to perform authorized work competently; but not to make important decisions for one. Auto mechanics cannot make repairs without authorization to do so. Most repair shops use order forms on which the work to be done is specified and then signed by the owner. If the mechanic discovers other problems requiring further significant repairs, further authorization must be obtained. However, if an owner trusts a mechanic, the latter may be authorized to check out the engine (or car) and do whatever is needed, provided it does not cost

---

[6]See Donald Oken, "What to Tell Cancer Patients: A Study of Medical Attitudes," *J. Amer. Med. Assn.* **175**, 1125 (1961).

[7]William D. Kelly and Stanley R. Friesen, "Do Cancer Patients Want to Be Told?", *Surgery* **27**, 822–826 (1950).

more than some specified amount, such as $50. Permitting a mechanic to make such decisions falls under the first reason for allowing others to make judgments on one's behalf, namely, that the matter is not significant enough to attend to oneself.

Analogously, physicians must have informed consent from patients.[8] A physician cannot legally or morally perform diagnostic tests or make therapeutic interventions without a patient's consent. Such consent must be informed, which means that the obligation of truthful disclosure of all materially relevant facts must be met before the consent is acceptable. An automobile mechanic should explain how bad the brakes or shocks, or whatever, are before seeking permission to repair them. When major repairs are indicated, owners may go to another mechanic for a second opinion, just as patients may want a second doctor's judgment about an elective operation. Finally, a mechanic may describe alternative procedures, such as tightening and relining the present brakes or putting in a new set. So also, a physician should explain to a patient alternative procedures, for example, surgery or radiation therapy for a malignancy.

Two of the reasons advanced against the analogy between the physician–patient and mechanic–owner relationships make client consent even more important in the medical context. As noted above, the range of client interests affected by physicians is often greater than that affected by mechanics, and medical science may not be value-free. The more values likely to be affected, the more important decisions become and the harder it is for physicians to judge as the patient would. Hence, there is more reason for allowing patients than owners to make decisions themselves. Thus, the weaker the analogy in these respects, the stronger the obligation for physicians than for mechanics to secure client consent.

An automobile owner may be dissatisfied with the work done by the mechanic: the engine may stall at stop lights after a tune-up, the hand brake may not hold, and so on. The owner has recourse for slipshod, careless, or incompetent work. Likewise, a patient may be dissatisfied with the ministrations of a physician. The pain may continue, the medicine may produce considerable nausea, the illness may persist. In either case, an owner or patient may sue for negligent work. With respect to negligence, some people may claim that the analogy between the mechanic–owner and physician–patient relationships breaks down. A physician will probably respond that a cure cannot be guaranteed, though a mechanic can of course guarantee repairs. There

---

[8]Cobbs vs Grant, 8 Cal. 3d 229, 502 P.2d 1, 11, 104 Cal. Rptr. 505 (1972).

is a complete and strongly confirmed scientific theory or theories to explain the functioning of automobiles and their subsystems or parts. However, there are no complete theories of the functioning of human beings, or many of their subsystems. Although there is, of course, considerable and growing scientific knowledge about the functioning of human beings and their organs, much remains still unknown. It has even been argued that medical knowledge is necessarily fallible.[9] Because of this difference in knowledge, a physician cannot guarantee the outcome of treatments, though a mechanic can. Hence, untoward results are conclusive evidence of negligence or incompetence on the part of a mechanic, but not of a physician.

Although there is considerable truth in these claims, the differences and their significance are easily exaggerated. First, even if there is a complete (or nearly complete) theory of the operation of automobiles and their parts, it does not follow that in all cases a repair can be guaranteed to fix the difficulty. A mechanic, like a physician, is often confronted with symptoms, such as a loss of lights. There are several possible causes for a failure of lights. I once had difficulty with headlights failing at night. The cause could have been the alternator, a short in the wiring, and so on. One mechanic attributed the difficulty to the alternator. However, as it later turned out, the problem was a new fan belt, which had stretched and slipped. Since there are multiple possible causes of symptoms, even a complete theory does not enable one to guarantee that a specific repair will effect a cure. At best, with allowance for defective spare parts, a mechanic can guarantee that what is repaired will work as it should. The mechanic of course cannot guarantee that the entire system or car will work as desired.

Second, the limits of medical knowledge do not, of course, imply that physicians cannot be negligent. What does follow is that the standards of due care will vary between physicians and mechanics. The standard of care to which mechanics should be held is firmly tied to results, for example, that parts be securely mounted and correctly installed. There is an absolute standard of care in the sense that there are proper protocols of installation, tuning, and so on that do not vary. The standard of care for physicians, however, is a relative one. They

[9]Samuel Gorovitz and Alasdair MacIntyre, "Toward a Theory of Medical Fallibility," *Hastings Center Report* 5, 13–23 (1975). Elsewhere I criticize their view: see Michael D. Bayles and Arthur Caplan, "Medical Fallibility and Malpractice," *J. Med. Phil.*, (Sept. 1978).

ought to conform to the standard of practice of other physicians in similar communities and specialities.[10]

Finally, neither the mechanic–owner nor the physician–patient relationship is isolated from the wider community. Consequently, there are limits on the devotion of mechanics and physicians to owners and patients. There are some laws and obligations on a mechanic for the benefit of the wider community. Legally, an odometer cannot be turned back unless it is necessary and a record made. A mechanic also has a responsibility to the wider public that automobiles be safe for operation when lives other than the owner's may be at risk. Physicians also have responsibilities to the wider public that may limit their relations to patients. An obvious area of such responsibility pertains to communicable diseases. The confidentiality of physician–patient communications may need to give way for notification to others of their exposure to venereal disease. Similarly, the potentially disabling and dangerous problems of airline pilots and other similarly responsible employees, such as heart defects, may justify notifying employers of their illnesses.

A comparison of the auto mechanic–owner and physician–patient relationships indicates that the latter is not as unique and difficult to analyze ethically as is sometimes thought. Further similarities and ethical principles could be developed. The dissimilarities generally support more stringent requirements on physicians to make truthful disclosures of all relevant information and to obtain patient consent. Although the physician–patient relation is not the sole basis for medical ethics, in many contexts it is illuminating to view physicians as body mechanics.

---

[10]Restatement (Second) of Torts, § 299A.

14

Physician as Body Mechanic—Patient as Scrap Metal: What's Wrong with the Analogy

Colleen D. Clements

Department of Pediatrics and Department of Philosophy, University of Rochester, Rochester, New York

I

It is not often that I find myself in such consistent disagreement with a presentation as I have with Michael D. Bayles' "Physician as Body Mechanic." My disagreement ranges from the basic infrastructure of the analogy to the heuristic value of using it. Unfortunately, I find the disagreement fundamental and total.

Looking at the infrastructure, I am troubled by what I see as important differences in the two relationships, so important as to render analogous comparison inaccurate and distortive of reality. The auto mechanic–customer (consumer) relationship is best represented by the following diagram (1):

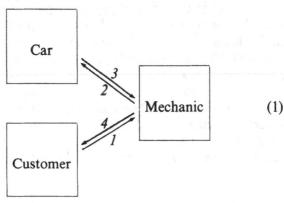

(1)

The customer contracts with the mechanic to fix whatever is wrong with the car. After working on it, the mechanic reports back to the customer. Manipulations on the car are *not* manipulations on the customer. What is at stake is not the existential integrity of the customer.

The physician–patient relationship can be represented by quite a different diagram (2), with one exception that I will discuss shortly:

$$
\boxed{\text{Patient}}
\begin{array}{c}
\xleftarrow{\ 4\ } \\
\xleftarrow{\ 3\ } \\
\xrightarrow{\ 2\ } \\
\xrightarrow{\ 1\ }
\end{array}
\boxed{\text{Physician}}
\qquad (2)
$$

Here, the patient presents her- or himself to a physician, who manipulates that patient in an effort to discover what the problem is and how best to treat it. Feedback from the patient helps inform the physician and allows a diagnosis and prognosis, which the physician then communicates to the patient in a way that he or she feels serves the patient's best interest. There is no separate "car entity" whose existence can be threatened without placing the customer's existence in jeopardy.

There are exceptions to this infrastructure diagram (2). These occur when a customer contracts with a physician to care for another human life for which the customer is responsible. [A parent takes a minor child to the pediatrician, or an adult offspring assumes decision-making authority for an unconscious parent, or a spouse assumes responsibility for an unconscious mate, or a sibling fills this role for a mentally incompetent sibling.] In any of these variations, the infrastructure can resemble that of the mechanic–customer relationship. But these are peculiar relationships and force modification of the physician–patient model both theoretically and practically. Whenever we have a "stand-in" for a conscious responsible patient, the medical model squeaks and acknowledges an aberration. Moreover, a child becomes the ward of the court or hospital because physicians perceive themselves as standing in the primary physician–patient role with the child as well as the parent, and in effect have two patients rather than one. The staff refuses to unplug the machinery keeping an unconscious (three flat EEG's) terminal elderly patient alive when the children wish heroic measures to cease because the staff perceives a dual responsibility: to the family as surrogate, but also to the elderly, dying patient. It may be possible to accommodate such variations into the infrastructure of the mechanic–customer relationship if we are determined to do so, and if we overlook the numerous instances when medical personnel practically assume the two-patient model or theoretically maintain it for the "stand-in" situations. I would prefer a two-patient diagram myself, as coming closest to representing all "stand-in" situations:

(2a)   (2b)

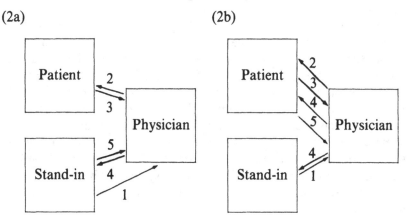

In diagram (2a) the stand-in presents the patient to the physician. The physician manipulates the patient and receives the necessary feedback to make the diagnosis and prognosis. Then the physician reports back to the stand-in, accepting the stand-in as decision-maker, informed-consenter, or prognosis-receiver. In diagram (2b) the same initial process occurs, but the physician acts as though the patient were the decision-maker, and does not accept the stand-in in that role.

I believe these diagrams more accurately portray the infrastructure of the examples, show up some major problems with maintaining Professor Bayles' analogy, and demonstrate rather clearly that the relationship between expert E and nonexpert N *crucially* depends on the nature of S, the subject matter. E must relate to S in ways determined by the nature of S, and N must relate to S either in reflexive identity, as a stand-in, or as independent party to the interaction. There is no separate E–N relationship. If N has no car, N does not relate to a car mechanic as repairman. In the medical model, N either is reflexively S, or is accepted as a stand-in for S, and/or S is a second patient.

Further, the humanity of social interactions is sufficiently at risk already in bureaucratic structures: the danger of making human contacts external relations cannot be overly stressed, but is especially important in medicine, the "humane science."[1] And there remains the

[1]David Mechanic, *The Growth of Bureaucratic Medicine: An Inquiry into the Dynamics of Patient Behavior and the Organisation of Medical Care,* Wiley, New York, 1976; Sander Kelman, "Toward the Political Economy of Medical Care," *Inquiry* **8**(3), 30–36 (1971). Jonathan A. Gallant and John W. Prothero. "Weight-Watching at the University: The Consequences of Growth," *Science* **175**, 381–388 (1972); Victor R. Fuchs, *Who Shall Live?*, Basic Books, New York, 1974, Chap. 3, and also, his Wilson

ethical problem of considering a human being in only external relationships: *should* we treat people as objects?

There are some further difficulties:

1. A car is not a homeostatic structure for the most part. It has
few negative feedback loops. I can name the thermostat mechanism as
one stasis subsystem, but am hard-pressed to think of others. As the
amount of gas continues to decrease, for example, there is no loop to
slow speed or otherwise reduce consumption. The car can continue at
55 mph until gas is totally consumed. The human body in contrast is
a marvelous stasis system, as Cannon perceived. A "body mechanic"
from the automotive world is simply not an accurate metaphor for a
physician, either theoretically or practically. We do not need philosophic distinctions that fragment the human system. What we do need
is a systems approach that can handle development and process
through time.[2]

2. Cars can be disposed of by their owners without involving any
loss of existential integrity, in fact they are routinely traded in. Although $5000 may represent a significant financial loss (the average
worst thing that could happen to a car owner as car owner), when N
is S the loss is total, it is the loss of N. This is the frightening bottom
line that all physicians, and no auto mechanics, must face. It is one of
the fundamental facts of the medical professions and is instrumental in
shaping all its relationships. Serious mistakes *are* buried.[3]

3. There is a philosophical adjustment we could make to fit the
mechanic–customer relationship into my proposed structural diagram
and rescue the analogy. I consider it a regressive move that could not
be pragmatically justified, but I will bring it to your attention because
I feel it brings to light one of the underlying philosophic dangers of the
analogy. We could conceive of the human body as separate from the
person or mind of the patient. We might then invite the return of the

Day Lecture, University of Rochester, October 12, 1977; Roy Branson, "The Seculari-
zation of American Medicine," *Hastings Center Studies* **1** (2) 18–28 (1973); Eric
Cassell, "Illness and Disease," *Hastings Center Report* **6**(2), 27–37 (1976); Michael
Jellinek, "Erosion of Patient Trust in Large Medical Centers," *Hastings Center Report*
6 (3), 16–19 (1976); Lewis Thomas, "Commentary: The Future Impact of Science and
Technology on Medicine," *BioScience* **24**(2), 103 (1974).

[2]Robert Kohn and Kerr White, eds., *Health Care: An International Study*, London, Oxford University Press, 1976; William Gray and Nicholas Rizzo eds., *Unity
Through Diversity*, New York, Gordon and Breach, 1973; Ernest Becker, *The Denial
of Death*, New York, Free Press, 1973.

[3]Eric Cassell, "Making and Escaping Moral Decisions," *Hastings Center Studies*
1(2) 53–62 (1973); Colleen D. Clements, "Death and Philosophic Diversions," *Phil.
Phenomenol. Res.* (forthcoming).

ghost to the machine. Real "humanity" would consist of the personality or mind (customer), which could be considered separate from the body (car) in crucial ways: the body could stop functioning, but the person or mind would continue to function: major damage to the body could occur without adversely effecting the person or mind. Although this suggestion might appeal to some, I find that three flat EEG's are very instructive and very persuasive. When and if proponents of "life after life" present me with after-death experiences that have occurred with no brain activity (three flat EEG's), then I would be more willing to consider a philosophic move back to the ghost in the machine.

4. Status in society tends to indicate the social value placed on a role. If the $E-N$ relationship could be constructed without reference to S, it would seem prima facie that similar $E-N$ relationships should receive correspondingly close social valuation. But the status gulf between automechanic and physician is apparent and enormous.

II

However, it is not only on the conceptual level that I cannot accept the proposed analogy. I think that there are major and compelling pragmatic reasons for avoiding the metaphor as well, because the use of it would adversely effect the attempt to solve current problems in medical ethics, and in fact it might contribute to the very problems that are slowly being identified. I want especially to emphasize the inadequacy of the body-mechanic model to deal with these current medical problems because any proposed model needs to demonstrate its fruitfulness in handling these perceived major concerns if it is to have any relevance for the medical community. For those of us who feel the pragmatic criterion of truth has value, the consequences of a model also determine its accuracy.

1. Patient as Object or as Person. Let me use a counselling case as an example.[4] A thirty-one year old married nurse with four previous pregnancies (two spontaneous abortions, one ectopic, one hydrocephalic stillbirth) requested counselling to determine the risk of another unsuccessful pregnancy. Actually, she was very upset, in tears, at the thought of risking another unsuccessful pregnancy. Her husband wanted to try again. No problems in the pedigree were indicated, toxoplasmosis and rubella were eliminated, and there was no data on

---

[4]Case studies taken from Genetics Clinics, Prenatal Detection Program, University of Rochester Medical Center, Richard Doherty, M.D., Program Director.

the hydrocephaly. The real reason for presenting to the physician is probably psychosocial: "assure us we won't have to go through the tragedy again; we're afraid." The physician tends to provide a release mechanism for the articulation of psychological and social problems.

2. *Patient as Stasis-System or Patient as "Phlebitis in 5A".* The patient is a much more complicated system than a car-model suggests. I can cite a chronic lymphocytic leukemia (CLL) case in which symptoms masked life-threatening intestinal blockage; standard CLL therapy was contraindicated in view of imminent surgery. There was also a pelvic bone cancer case in which treatment with female hormone resulted in impotency for the husband, whose self-image proved to be crucially tied to the frequency and success of intercourse. Combined with long-standing psychological problems, this therapy ultimately resulted in a murder–suicide.

3. *Death with Dignity.* The "physician as body mechanic" metaphor divests death of its profound significance and trivializes the human situation. At this time, medicine is finally beginning to deal with the complexity of terminal illness. But this analogy represents a death-denial move; cars do not die, they can be traded in, indiscriminately cannibalized for parts, and so on. Dr. Balfour Mount's hospice experience is not understandable in terms of a body-mechanic model.[5]

4. *Right to Health Care in a Resource-Scarce World.* The body mechanic would prejudge this issue in medical ethics. To talk about the "right to car maintenance and running condition" makes little sense, since car ownership is determined by the financial status of the owner and repairs are determined by the ability to pay for them. When we talk about the programmatic break-even cost of amniocentesis at the Prenatal Detection Center, we are using a much more complex economic model than the individual car-owner and maintenance model.[6]

---

[5]Dr. Mount is Director of the Palliative Care Unit, Royal Victoria Hospital, Montreal. See Cassell, *op. cit;* Clements, *op. cit.,* and "Pain and Terminal Illness," University of Rochester Symposium, March 1977, and "The Child and Death," University of Rochester Symposium, Sept. 1977.

[6]Fuchs, *op. cit.,;* Clark C. Havighurst and James Blumstein, "Coping with Quality/Cost Trade-offs in Medicare Care: The Role of PSRO's," *Northwestern University Law Rev.* 70 (March, April 1975), pp. 6–68; Andreas G. Schneider and Joanne B. Stern, "Health Maintenance Organizations and the Poor: Problems and Prospects," *Northwestern University Law Rev.* 70 (March, April, 1975), pp. 90–138; William F. May, "Code, Covenant, Contract, or Philanthropy," *Hastings Center Report* 5(6), 29–38 (1975); Roger D. Masters, "Is Contract an Adequate Basis for Medical Ethics?," *Hastings Center Report* 5(6), 24–28 (1975); Gordon Bermant, Peter Brown, and Gerald Dworkin, "Of Morals, Markets and Medicine," *Hastings Center Report* 5(1), 14–16 (1975); David Mechanic, "Rationing Health Care: Public Policy and the Medical Marketplace," *Hastings Center Report* 6(1), 34–39 (1976).

5. Primary Relationship or Bureaucratic Medicine. The body mechanic would overlook the personal doctor–patient relationship; the car does not care who works on it. I can only recall the eloquent plea of a father with thalessemia major-Lepore anemia that his seven-year-old daughter with the same condition be assigned one consistent doctor at the Medical Center to give her the required transfusions every three weeks. The child was becoming very upset and needed to relate to some particular provider, especially when her last transfusion had to be started nine times. But the need to relate to another human being is not only a childhood need. The car model runs the risk of preventing us from fully realizing the costs of bureaucratic medicine, and we need bear all the costs in mind when considering the cost–benefit analysis of health-care delivery systems such as HMO's or National Health Insurance.[7]

6. Self-Regulation and Medicine as a Profession. The model also appears to me to prejudge the self-regulation issue. We know auto mechanics are not self-regulated and, if the analogy holds, physician body-mechanics also should not be self-regulated. Since Friedson isolates such autonomy as crucial to professionalism, it would pay to proceed cautiously and not use a model that has already decided the issue.[8] The question of how much autonomy is necessary for the continued professional growth of medicine is part of a murky problem that includes questions of lay consultation in health policies, government regulation of research, academic and scientific freedom of inquiry, and so on. The proposed analogy cannot aid us here since it has prematurely decided the issue.

My conclusion, then, is precisely the opposite of that of Professor Bayles. A comparison of the mechanic–customer and physician–patient relationship does indicate that the latter is unique in important respects and that philosophers had best proceed very cautiously in the analysis of biomedical ethical issues. In fact, I feel we had best attribute significant importance to the clinician perspective, especially since philosophers already tend to have an excessively rationalistic model of human intercourse.

---

[7]David Mechanic, *The Growth of Bureaucratic Medicine,* New York, Wiley, 1976; Benjamin B. Page, "Socialism, Health Care and Medical Ethics: A Letter from Czechoslovakia," *Hastings Center Report* 6(5), 20–23 (1976).

[8]Eliot Freidson, *Profession of Medicine,* New York, Dodd, Mead, 1970; also *Professional Dominance,* New York, Atherton Press, 1972.

15

Internal or External Physician—Patient Relationships, A Response to Clements

Michael D. Bayles

Department of Philosophy, University of Kentucky, Lexington, Kentucky

In her thoughtful discussion, Professor Clements exaggerates our disagreement by considering topics I did not discuss. I confined my discussion to relations between individual primary-care physicians and individual patients without a life-threatening disease. Consequently, I made no claims about, nor do I wish to here pursue, what she calls "stand-in situations," patients who are terminally ill or have seriously life-threatening diseases, policy analysis for the allocation of health resources, or professional self-regulation with respect to health policy, research, or academic freedom.

The fundamental difference between Professor Clements and me regards whether the physician–patient relationship is external or internal. Internal relations affect and depend upon the nature of the terms related, so that a difference in one makes a difference in the other. External relations, such as "standing next to," are not so connected to their terms. Not everything may be a term of a relation. Only objects in space may be next to one another.

Professor Clements appears to believe that the physician–patient relationship, or any contractual relationship, is internal. She claims that "the relationship between expert E and nonexpert N *crucially* depends on the nature of [the subject] S. E must relate to S in ways determined by the nature of S. . . . There is no separate E–N relationship."[1] However, I do not find her arguments for this claim convincing. *1.* The crucial issue is her suggestion that N and S are identical and that the analogy implies a ghost-in-the-machine conception. In my paper, I admitted that the relationship between a person and that individual's body is not one of ownership, and thus differs from the relationship an owner has to an automobile. But this fact does not entail that the $E-N$ relationship is different. Although I do not own my daughter, it does not follow that my relation to her private music teacher is any different from that in other personal service contracts. Moreover, I do not think the person–body relationship is one of identity such that any change in the body entails a change in the person. All that one needs for the analogy is a conceptual difference between the subject (body) and the person, which there is. Also, overidentification with one's own body may be disastrous, as is indicated by Professor Clements' example of murder and suicide owing to impotency that resulted from a specific cancer treatment.

Her other arguments are less plausible. 2. She asserts that if a person has no car, that individual does not relate to a car mechanic as a repairman. This shows only that there is no relationship when there is no subject for it; an individual who does not want personal health information does not relate to a doctor as a physician, as I am now not relating to Dr. de Verber (the moderator) as a physician. 3. She asserts that the body is mostly a homeostatic structure, but a car is not. However, one's relationships to car and computer repairmen do not differ simply because computers have many feedback loops and homeostatic structures.

Professor Clements' pragmatic reasons for rejecting the analogy are even less convincing. Of course, that a view has undesirable consequences shows nothing about its correctness or aptness. In any event, the analogy does not have the undesirable consequences she depicts, and it has better consequences than the view she seems to adopt. Although she never explicitly so states, Professor Clements appears to adopt a paternalistic model of the physician–patient relationship. She writes that the physician "manipulates" the patient to discover information and "communicates [information] to the patient in a way he or she feels serves the patient's best interest."[2] On the fiduciary model I adopt, the physician makes a full, truthful disclosure of information and the patients determine their own best interests. I submit that on this model the patient is treated more as a

---

[1]Colleen D. Clements, "Physician as Body Mechanic, Patient as Scrap Metal: What's Wrong with the Analogy," this volume p. 179.

[2]*Ibid.*, p. 180.

complex person and less as an object than on the paternalistic, doctor-knows-best, model.

Finally, Professor Clements and most physicians seem to believe patients come to them more for psychological reassurance than treatment and cure. I have heard a group of physicians maintain this position in the face of its denial by a group of lay people. Even if true, the body mechanic analogy and fiduciary model do not seriously prevent personal reassurance. Obtaining information is what usually provides reassurance—learning that the swelling is not a tumor, the rash is minor, and so on. Thus, by full, truthful disclosure a physician will also usually provide personal reassurance. Moreover, an auto mechanic also often provides such reassurance—the rattle in the front end is just a piece of loose metal, not an engine problem.

IV

INFORMED CONSENT
AND PATERNALISM

16

The Ethical Content of Legally Informed Consent

Bernard M. Dickens

Faculty of Law, University of Toronto, Toronto, Canada

I. INTRODUCTION

The legal approach to consent always has been to regard it as a matter of substance, and not merely of form. In this sense, the legal quality of consent has coincided with an ethical function in recognizing individual autonomy. A person's signature on a contract, will, hospital consent form, or other document is not taken as conclusive proof that the person is in fact a consenting party to the transaction. The presence of an authenticated signature creates a presumption of consent, but the signatory and others who may hope to profit from showing the signatory not to be bound remain free to demonstrate, despite the signed document, that the signatory was not a consenting party in law to the transaction the document evidences or purports to create.[1]

Several grounds may be invoked to nullify the apparent consent. Mental incompetence[2] and minority age[3] may negate apparent consent on the basis of the signatory's own incapacity (although both mental incompetents[4] and minors[5] may remain bound by certain transactions,

---

[1] *Halushka* vs *University of Saskatchewan,* (1965), 53 D.L.R. (2d) 436.

[2] See generally J.A. Coutts, "Contracts of Mental Incompetents," *Special Lectures of the Law Society of Upper Canada,* 49–75 (1963).

[3] See generally W. Wadlington, "Minors and Health Care: The Age of Consent," *Osgoode Hall Law J.* **11,** 115–125 (1973); A. Shaw, "Dilemmas of 'Informed Consent' in Children," *New England J. Med.* **289,** 885–890 (1973).

[4] *Robertson* vs *Kelly,* (1883), 2 O.R. 163.

[5] *Booth* vs *Toronto General Hospital,* (1910), 17 O.W.R. 118; *Johnston* vs *Wellesley Hospital,* (1970), 17 D.L.R. (3d) 139.

notably those designed for their benefit). Vitiating factors introduced by the other side to the transaction might include fraud, duress, and undue influence, all of which tend to induce a form of consent that lacks its essential legal substance.

The concept of undue influence arises from legal relationships based not on the competitive principle of the marketplace, where the rule *caveat emptor* (let the purchaser beware) applies, but on the duty of trust, that is, the duty to observe the utmost good faith because of inequality of parties.[6] Inequality, or inequity, is found when one party must of necessity repose trust in the other. These special trust relationships, called fiduciary relationships, include for instance the doctor–patient and the lawyer–client relationships. They require the party delivering a service to make an adequate disclosure of information within the special knowledge of that party not possessed by the other. In this way, the doctor–patient relationship is founded in law upon conscience, rather than upon self-seeking commerce, and presupposes that the physician knows the patient well enough to cater to that patient's particular requirements.[7]

Consent is regarded as a quality that arises in human relationships, and its presence is measured by reference to applicable legal norms. The degree of information possessed by each party accordingly is required to be appropriate to the nature of the interpersonal transaction. In impersonal commerce, each party must be entirely self-reliant, and will be bound by a poor bargain unless misled by false statements or misrepresentations made by or on behalf of the other party. The norm is negative, consisting in not wilfully misleading, and one party has no duty to protect the other from its own self-deception and ignorance. In fiduciary relations, however, such as those between doctor and patient, the norm is positive, and the stronger side, meaning the side with the power of knowledge, is required to ensure that the partner to the transaction has adequate information and is not influenced by misunderstanding or impaired by ignorance.[8] The fiduciary must ensure that the quality of the partner's consent is adequately informed. It need scarcely be added that, as in the relationship between equals, fraud or duress destroys the ties of mutuality.

---

[6] *Re C.M.G.*, [1970] Ch. 574; *Vanzant* vs *Coates*, (1918), 39 D.L.R. 485; *Natanson* vs *Kline*, (1960), 350 P. 2d. 1093.

[7] On the standard of care required, see A.M. Linden, "The Negligent Doctor," *Osgoode Hall Law J.* **11**, 31–39 (1973).

[8] *Halushka* vs *University of Saskatchewan, op. cit.*

II. THE LEGAL DYNAMICS OF INFORMED CONSENT

The development of the legal concept of informed consent has received considerable impetus in recent years through the dynamics of medical malpractice litigation.[9] A person's bodily integrity entitles that individual to be free of the interference of another, except when consent to the other's bodily invasion has been earlier given. Wilfully inflicting unauthorized interference constitutes the civil wrong of battery,[10] and inadvertent or careless interference arises from negligence, so that classically the party suffering the invasion might frame the action for legal redress of an injury or insult in the context of either battery or negligence law.

There are four legal elements to a negligence action. The first is the existence of a legal duty of care, the second is a breach of that duty, the third is damage, and the fourth is causation, meaning that the damage must be shown to result from the breach of the duty of care.

1. A legal duty of care may be assumed voluntarily by parties to a contract,[11] but in principle the law imposes a duty of care[12] upon everyone whose conduct may affect another person within reasonable foresight (even if that person is not yet born or conceived). A duty of care must therefore be established to the satisfaction of the court before which an injured party claims damages.

2. Breach of duty of care is essentially a matter of fact, but insofar as it concerns the defendant to the action falling below a legally acknowledged standard of care, it involves a legal component. This is especially so in medical matters, since conformity to the local medical standard may nevertheless be legally inadequate; the law may require a higher standard of care than general local practice satisfies.[13]

3. The existence of damage is also in principle a matter of fact, but even here the law has principles of classification. The law

---

[9]See generally D.A. Geekie, "The Crisis in Medical Malpractice: Will It Spread to Canada?" *Can. Med. Assn. J.* **113,** 327–334 (1975).

[10]*Mulloy* vs *Hop Sang,* [1935] 1 W.W.R. 714.

[11]A contract exists even when payment on behalf of one party is made by a third party, such as a provincial or private health insurance program. On government financed medical care, see S. Andreopoulos (ed.) *National Health Insurance: Can We Learn from Canada?,* Wiley, New York (1975).

[12]For a physician's duty of care irrespective of contract, see *Everard* vs *Hopkins,* (1615), 80 Eng. Rep. 1164.

[13]See A.M. Linden, *op. cit.,* p. 32, and *Helling* vs *Carey,* (1974), 519 P. 2d 981.

has long been reluctant to acknowledge as a form of legal damage, for instance, the birth of a normal healthy child,[14] and actions for improper performance of contraceptive protection, sterilization, or vasectomy have had to overcome the law's esthetic disfavor.[15]

4. Causation, aimed at showing the damage to be not simply subsequent to, but consequent upon, breach of the duty of care, usually requires expert medical testimony. Legal liability attaches not just to being negligent in itself, but to causing damage by being negligent. If the injury or death complained of would have occurred in any event, the legal action for negligence will therefore fail.[16]

Because of these legal, factual, and medical obstacles to be surmounted in negligence actions, a number of litigants prefer to sue for battery, which consists simply in unauthorized touching. This can mean touching with no appearance of consent, but also touching with no substance of consent, because the consent given, perhaps in a consent form, is ineffective. This may be because the medical defendant failed to give adequate information to render the consent legally operative.

The advantage of the action for battery, or, as it is sometimes called, surgical assault, is that the plaintiff need not show or allege any lack of care or skill on the part of the physician concerned,[17] and such a suit can succeed without proving that the battery or assault caused any specific injury.[18] The injury shown to have resulted from the wrong will affect the amount of damages recovered, of course, but success in the litigation, which in Canadian jurisdictions includes recovery of legal and other costs from the unsuccessful party, does not depend on proof

---

[14]See generally *Zepeda* vs *Zepeda*, (1963), 190 N.E. 2d 849. For an unsuccessful action by a severely disabled child, see *Gleitman* vs *Cosgrove*, (1967), 227 A. 2d 689, but for a successful claim, see *Park* vs *Chessin*, (1977), 400 N.Y.S. 2d 110.

[15]See generally, H. Krever, "Some Legal Implications of Advances in Human Genetics," *Can. J. Genet. Cytol.* 17, 283–296 (1975); and B.M. Dickens, "Eugenic Recognition in Canadian Law," *Osgoode Hall Law J.* 13, 547–577 (1975). The position in the United States is more accommodating of claims; see J.S. Ranous and J.J. Sherrin, "Busting the Blessing Balloon: Liability for the Birth of an Unplanned Child," *Albany Law Rev.* 39, 221–251 (1975).

[16]*Bolam* vs *Friern Hospital Committee*, [1957] 2 All E.R. 118.

[17]It is obvious that an assault is not defensible by proof that it was skilfully performed.

[18]*Mulloy* vs *Hop Sang, op. cit.*

of causation, but simply requires proof of touching without consent: for instance, when the consent apparently given, and under which the physician claimed to act, was not an informed consent.

It may be added that a hybrid form of action is now developing, in which the physician's failure adequately to inform the patient before touching is alleged to be a breach of the physician's duty of care, and the wrong is treated as negligence.[19] Causation then requires the plaintiff to show that had proper information been provided, for instance, of the risks and possible side-effects, consent to the particular medical procedure would not have been granted, and therefore the injury it caused would not have been suffered. This overcomes the defendant's argument that the procedure bears an irreducible minimum risk of injury that no amount of care and skill can eliminate, and enables a plaintiff to succeed in proving negligence without having to rely upon medical expert witnesses.

III. THE ELEMENTS OF INFORMED CONSENT

Whether the complaint is of surgical assault or negligence, therefore, the plaintiff alleges lack of informed consent, leading us to ask what knowledge renders consent informed. The traditional elements are five in number. The patient must be informed of:

1. The prognosis if one remains untreated. It must be made clear that even when there is only a single "treatment of choice," the patient still retains a choice, namely not to take treatment. The patient must therefore be told what the future is likely to be if treatment is declined. Thus one's refusal of treatment must be as informed as one's consent to treatment.
2. The treatment alternatives, when they exist.[20]
3. The discomforts and risks attached to each option, including the risk of a treatment alternative being unsuccessful. The

---

[19]See L.E. Rozovsky, "Consent to Treatment," *Osgood Hall Law J.* **11**, 103 (1973); L.B. Jaeckel, "New Trends in Informed Consent," *Nebraska Law Rev.* **53**, 66–92 (1975); and A. Meisel, "The Expansion of Liability for Medical Accidents: From Negligence to Strict Liability By Way of Informed Consent," *Nebraska Law Rev.* **56**, 51–152 (1977). See also the discussions in *Kelly* vs *Hazlett* (1976), 75 D.L.R.(3d) 536 and *Reibl* vs *Hughes,* (1977), 78 D.L.R. (3d) 35 (appeal pending).

[20]Whether information may be limited to alternatives available within a given facility, or must include treatments available elsewhere, may depend upon the patient's means to go elsewhere.

physician is not taken to guarantee the success of any treatment,[21] unless preceded by the physician's express undertaking.[22]

4. The possible side-effects of each treatment, even when it is successful.

5. The purposes and potential benefits of each treatment.

The physician should also offer to answer any questions, respond to them fully and honestly,[23] and tell the patient that, even if a procedure is commenced, it may be freely discontinued without prejudice to the patient's entitlement to future treatment.

Beyond this, physicians are entitled, and perhaps bound, to put on record their own views regarding the most suitable treatment, but the decisive choices must remain those of the patients, and not the physicians. The physicians must therefore not overstate the benefits of the treatments favored, although they may state as facts which treatments they consider preferable.

Once treatment is commenced, furthermore, physicians must keep their patients' informed consent up to date. Informed consent is not just instrumental at the time of forging the therapeutic doctor–patient relationship, but remains a principal condition throughout its endurance. Thus, if an agreed-upon treatment becomes discredited owing, for instance, to the patient's unfavorable response or to developments in the evolving medical literature, the physician must keep the patient adequately aware of these developments and their implications.[24] Equally, if a new treatment becomes current in developing medical usage, the physician must inform the patient if it becomes an available therapeutic option. The physician's duty is not to keep the patient apprised of each medical shift, but to maintain a sufficient degree of professional awareness to interpret relevant changes to the patient, and to prevent the patient from being locked into an inferior or less advantageous treatment by the physician's ignorance, and therefore by the patient's own.[25]

---

[21]*Hughston* vs *Jost,* [1943] O.W.N. 3.

[22]See *Town* vs *Archer,* (1902), 4 O.L.R. 383 at 388 (*per* Falconbridge, C.J.); *Noel* vs *Proud,* (1961), 367 P. 2d 61; *Guilmet* vs *Campbell,* (1971), 188 N.W. 2d 601.

[23]See *Smith* vs *Auckland Hospital Board,* [1965] N.Z.L.R. 191.

[24]See B.M. Dickens, "Information for Consent in Human Experimentation," *University of Toronto Law J.* **24,** 381–410, at 404 (1974).

[25]On the physician's duty to keep up to date, see *Ostrowski* vs *Lotto,* [1969] 1 O.R. 341 at 355 (*per* Keith, J.) (reversed on another point, [1971] 1 O.R. 372, affirmed (1972), 31 D.L.R. (3d) 715).

IV. THE STANDARD OF INFORMED CONSENT

The law does not render the physician's duty of disclosure absolute. United States' courts have gone to the extent of quantifying the level of disclosure on an objective scale, and have ruled, for instance, that a one per cent risk of injury should be disclosed to the patient,[26] although the risk, for instance, of contracting serum hepatitis (hepatitis B) from routine blood transfusion, which has been estimated at 0.013%, need not be disclosed.[27] Canadian courts prefer a more subjectively assessed standard, and have, for instance, exonerated a physician who failed to disclose a 20–30% risk of hearing loss when the particular patient, who experienced the hearing loss, could not have made any intelligent choice between what to that patient would have been bewildering medical alternatives.[28]

Paradoxically, the Canadian subjective test of disclosure is probably more demanding than the American objective test, and accords more fully to the ethical requirement of regarding the patient as a particular individual with personal characteristics and needs. The physician must give the patient such information as that patient will require to exercise a reasonable choice among treatment alternatives. The standard is not what information a reasonable patient would require, but what information this particular patient requires.[29]

It follows that if the patient cannot cope with information, such as in the hearing loss case, it need not be provided.[30] It also follows that if information that the usual patient can accept will cause the particular patient undue distress to learn, it also should not be provided. Medical malpractice can consist in overinforming as well as in underinforming if, for instance, the additional information causes the patient irrationally to reject a reasonable treatment or treatment alternative. The physician must walk a narrow line between underinforming and overinforming, and can discharge this legal responsibility only by adequately knowing the patient as an individual person, which is of course the physician's ethical duty.

The right to withhold dysfunctional information is called the physician's "therapeutic privilege." The physician need not explain in detail, for instance, the actual medical techniques to be used or the phar-

---

[26]*Canterbury* vs *Spence,* (1972), 464 F. 2d 772.

[27]*Sawyer* vs *Methodist Hospital of Memphis,* (1975), 552 F. 2d 1102.

[28]*Male* vs *Hopmans,* [1967] 2 O.R. 457.

[29]*ibid.,* p. 465.

[30]*ibid.,* and see *Natanson* vs *Kline, op. cit.*

macological operation of a drug, as long as the general nature of the treatment or drug is fully understood. In addition, the irreducible dangers inseparable from any operation, such as death under an anesthetic, the danger of infection, or of gas gangrene or gangrene, need not be stated, in the same way that the risk of contracting serum hepatitis from routine blood transfusion need not be stated.[31]

Beyond these standard nondisclosures, the physician may also choose to withhold information that might counteract an available therapy or induce a harmful effect, although the burden is always upon that physician to justify the prediction of the patient's likely idiosyncratic response that permits this selective nondisclosure; that is, all normally provided information is withheld at the physician's own risk. The reasonable fear of inadvertently inducing a side effect in a particularly suggestible patient by mentioning the possibility that it might occur may, for instance, justify nondisclosure.

V. THE FUNCTIONS OF INFORMED CONSENT

Most medical treatment is given upon the basis of a legal contract, even if it is generally unwritten and payment is made indirectly, through a provincial or private health insurance plan.[32] It has been seen that the fiduciary or equitable nature of the doctor–patient relationship requires the doctor to ensure that the patient has adequate information to exercise an independent will in entering and participating in the contractual relationship.[33] Although the contractual setting may appear exclusively legalistic, it may also provide a basis for the promotion of ethical standards, because it serves the significant personal nonlegal interests of both contracting parties.

The major ethical function of informed consent is to enhance the patient's individual autonomy. It redresses the nonprofessional's scientific disadvantage by permitting the formation of an intelligent opinion, allowing the patient genuinely to agree to what may be undergone and to the risks that may be run. This preserves for the patient not only intellectual dignity, but also bodily integrity.

The service to the patient's dignity and integrity protects his or her status as a human being. Before the patient's body may be subjected to treatments, there must be intellectual commitment to what the treat-

---

[31]See *Sawyer* v. *Methodist Hospital of Memphis, op. cit.*
[32]See note 11 above.
[33]See note 6 above.

ment entails. The patient is not simply informed, moreover, but must be actively involved in the decision-making process as it affects his or her medical, and perhaps human, destiny; that is, the patient possesses influence and a measure of self-determination over her or his personal future, and is not entirely dependent upon the will of others. The patient engages in the medical enterprise as a partner, and not simply as a client, subject, or potential victim. His or her decisions are no less rational than those of the other participants, who remain accountable to the patient for their actions and motives throughout their involvement with that patient.

The obligation to provide adequate information for consent compels physicians to recognize when they may be pushing towards the frontiers of knowledge, and when they are extending or departing from regular medical practice.[34] This is not, of course, to condemn or necessarily inhibit experimentation or therapeutic innovation. Indeed, a medical profession that did not experiment or innovate would be liable to ethical condemnation. This does require physicians, however, to acknowledge to themselves no less than to patients that they are entering untested areas or using untried methods. They must inform patients of all experimental or innovative aspects of proposed treatment,[35] and engage them in respectful voluntary partnerships. At the frontiers of knowledge, physicians' duties may be to inform patients not only of what is known, but also of what is not known. Thus, the need to obtain informed consent may enhance the scientific quality and validity of treatment, and encourage self-scrutiny within the medical profession.

VI. THE EFFECT OF INFORMED CONSENT

The legal effect of informed consent provides a strong reason why physicians might be enthusiastic to pursue it; informed consent shifts the risk of a medical procedure onto the patient. Autonomy and self-determination are associated with the concept of individual responsibility, and the autonomously acting patient personally assumes the risk of non-negligently caused injury. Law and ethics coincide in finding that when a person aware of an inherent risk freely agrees to undertake it,

---

[34]See Dickens, "Information for Consent in Human Experimentation," *op. cit.*, and "Contractual Aspects of Human Medical Experimentation," *University of Toronto Law J.* **25,** 406–438 (1975).

[35]See B.M. Dickens, "What is a Medical Experiment?" *Can. Med. Ass. J.* **113,** 635–639 (1975).

he or she is responsible for any injury suffered, and has no claim against the instrument or agent of the injury. The "voluntary assumption of risk" doctrine expresses a philosophy of individualism, and absolves a non-negligent physician from liability for injury that the patient suffers after giving free and informed consent to a medical procedure for which the risk was known to be an inseparable part.

It has been seen that the mere absence of negligence may not absolve the physician from legal liability, since there may still be liability in battery for unauthorized touching, meaning touching without (informed) consent.[36] Once legal liability is established, moreover, whether in negligence or battery, it may run to the full extent of any injury the patient suffered that was reasonably foreseeable, however unlikely it may have been. This mundane consequence may reinforce the physician's ethical incentive to ensure that the consent to treatment given by a patient is adequately informed. The law not only requires ethical practice, but rewards it.

VII. DILEMMAS OF INFORMED CONSENT

The extent to which the patient's informed consent to treatment protects the physician who administers it raises one of its ethical dilemmas, that is, the dilemma of whether the physician seeking informed consent should tell the patient that consent to medical treatment is granted at that patient's own risk. This might usefully encourage the patient, for instance, to arrange adequate insurance coverage for self and family alike. However, since bearing the risk is an unavoidable consequence of law that the patient is presumed to know, the physician does not need to add this legal information to the medical information that furnishes the substance of consent.

Ethical physicians must nevertheless ask themselves whether informed consent is sought to protect their own interests or those of patients because the latter may require doctors to provide more information than is necessary for the physicians' legal defenses. If, for instance, a treatment bears the risk of causing impotence or sterility, this may affect the patient's sexual partner and, in time, the stability of their family and domestic life.[37] The non-negligent physician will not be legally responsible, of course, for the patient's future distress in personal

---

[36]See *Mulloy* vs *Hop Sang, op. cit.*
[37]See generally I.N. Perr, "Legal Aspects of Sexual Therapies," *J. Legal Med.* **3** (1), 33–38 (1975).

relationships, but may ethically be expected to have had the foresight and care to direct the patient's thought to these wider and perhaps more sensitive considerations before possibly irreversible treatment has begun.

An effect of obliging the patient to entertain unwelcome and depressing thoughts may be that the patient declines to undertake an objectively advisable treatment, perhaps acting on emotional rather than rational grounds. Informed consent to treatment includes the concept of informed consent to decline treatment, of course, but this raises the dilemmas of malpractice by overinformation, and of informed consent becoming dysfunctional in reducing the individual's capacity for rational choice. The patient has no less a human right to his or her emotions than to rationality, of course, and in expressing the ethical function of informed consent as serving only rationality, we may be too constricting, and giving effect to a conceited intellectual preference for rationality at the cost of the patient's right not only to think, but to feel.[38]

The emotional component of the patient's choice can be negative or positive in effect. It can cause him or her to reject involvement in a procedure, but also to be irrationally committed to participation in a course of treatment that has few favorable health consequences to offer even after the physician has provided grounds for not entertaining unduly raised expectations. The attachment patients can develop to unproven or quack remedies is well known, but in reputable medicine, too, and especially in experimentation, a patient can develop an enthusiasm to be used. The physician–investigator must be prepared to keep this tendency in check, even to the extent of finding the subject unsuitable for a given procedure. The medical research community has norms to identify risks too great for even the fully informed subject to be allowed to run, and the physician–investigator must conform to such professional opinion even when the effect will be to deny the subject autonomy of choice. This arises when the patient's choice to take the risk is in conflict with the investigator's choice of action in what we have seen is a voluntary partnership, that is, a collaboration between two or more participants equally entitled to autonomy.

A further dilemma of informed consent concerns those patients who do not want to know: those patients relieving themselves of responsibility for choice by placing themselves unquestioningly at the disposal

---

[38]See S.I. Shuman, "The Right to be Unhealthy," *Wayne State Law Rev.* 22, 61–85 (1975), and the discussion in J. Jacob, "The Right of the Mental Patient to his Psychosis," *Modern Law Rev.* 39, 17–42 (1976).

of the physician. Physicians may suppose this attitude to exist in their patients rather more commonly than in fact it does, but, irrespective of its frequency, the attitude raises troublesome questions in principle. How far may a physician compel a patient to be more autonomous than that patient seems willing to be? How far can the physician go to counter the patient's preference or need for dependency, and to oblige him or her to be self-reliant and accountable for her or his own destiny, especially at a time when the patient's physical and psychological resources may be reduced? On the other hand, can the ethical physician uncritically accept the patient's surrender of autonomy, and passively collaborate in the patient's self-reification?

Further dilemmas concern the capacity for informed consent of legal minors, mental incompetents and, for instance, the senile. The legal approach to such dilemmas is to see them as a matter of process as well as of substance. The physician may be expected to reinforce or replace any consent a disadvantaged patient may give by seeking consent from a parent, say, or some other suitable relative or social guardian.[39] If this is not possible, the doctor should at least ensure that unimpaired individuals with a personal (as opposed to a purely bureaucratic) concern or affinity for the patient are made aware of all of the circumstances, and are free to enter objections or make representations for prior consideration.

A more acute and introverted aspect of this dilemma concerns those patients who are in law competent to act for themselves, but whose perceptions and capacity for thought have been reduced as a result of drug or other medical treatments administered as part of routine management. The ethically alert physician may want to know whether an action can ethically be based upon a reduced level of informed consent from the patient if that condition has in fact been induced by the physician, however therapeutically intended, especially when the physician's self-defensive alternative may be to jeopardize the patient's confidentiality and autonomy by involving others in decisions affecting treatment.

---

[39]See notes 2 and 3 above.

Comments on "The Ethical Content of Legally Informed Consent"

Abbyann Lynch

Department of Philosophy, St. Michael's College,
University of Toronto, Toronto, Canada

Professor Dicken's provocative paper merits extensive circulation and comment. I shall focus on three points likely to cause confusion, if not controversy, in the conceptual as well as the clinical area.

I. PERENNIAL PARADOX: "THE ETHICAL NATURE OF THE LEGAL"

The author's title* seems to presume some identity of ethics and law in the area of informed consent. There, and throughout the paper, a fundamental question remains unanswered: what is the basis for the distinction/relation between law and ethics in this (or any other) area?

Following a description of the legal requirements for "informed consent",[1] Professor Dickens allows some unclarified difference between law and ethics,[2] moving ultimately to a presentation of dilemmas[3] that underscore that difference. In demonstrating ambivalence throughout the paper, the author fulfills the promise of his title.

---

*At the time at which these comments were written, the title of Professor Dicken's paper was "The Ethical Nature of Legally Informed Consent."—*Eds.*

[1]Sections 1–3.
[2]Sections 4–6.
[3]Section 7.

Assuming an acquaintance with the more notable literature on the relations of morals and law, three choices in the matter seem open.

1. At certain points, Professor Dickens has assumed an identification of law and ethics without qualification. As a generalization, this appears quite inadmissible. It is tautological in concept, making further discussion difficult, and it is somehow "unworkable" as Professor Dickens' conclusion testifies. It could be argued that the dilemmas presented are exceptions to the "rule" of identification: given the many exceptions made, an attempt to formulate a more generally acceptable rule covering such exceptions must be made before the merit of the proposed generalized "rule" is allowed.

2. Though some would dismiss the case for so doing in terms of "fatal consequence",[4] Professor Dickens' dilemmas might well be used as the basis for an argument for the total divorce of ethics and law. Perhaps the difference suggested strongly at the end of this paper is the beginning of thought and action in this matter. If so, the apparent identification already made must be explained to avoid inconsistency.

3. As an alternative solution, it seems quite undeniable that the content of the law and ethics may on occasion be remarkably similar. One could reasonably expect, then, that the conclusions of the two would on occasion be similar, if not identical. This does not preclude the possibility (or truth) that law and ethics have differing grounds, goals, methods, and sanctions. "Coincidence" and "noncoincidence" between the two are thus explainable. If this is Professor Dickens' view, then it, too, requires further clarification in terms of the essay's title and content.

Setting aside the predicament imposed by the aforementioned paradox, Professor Dickens' paper presents a welcome opportunity for discussion of a similar difficulty in the matter of medical education and practice. The existence of (legal) "medical torture"[5] cannot be denied; some ambivalence about the "ethical" in the assessment of human experimentation protocols by University "Ethical" Review Committees is evident;[6] a growing "ethical cynicism" among health-care students has been suggested[7] even as the literature indicates the growth of preoc-

---

[4] *The Queen* vs *Dudley and Stephens,* (1884), 14 Q.B.D. 273 at 287.

[5] L. Sagan and A. Jonsen, "Medical Ethics and Torture," *New England J. Med.* **294,** 1427–1430 (1976).

[6] B. Barber, "The Ethics of Experimentation with Human Subjects," *Sci. Amer.* **234,** 25–31 (1976); S. Dymond, "The Strengths and Weaknesses of University Review Committees to Protect Human Rights," *Ann. Roy. Coll. Surgeons Eng.* **55,** 59–61 (1974).

[7] R. Morris and B. Sherlock, "Decline of Ethics and the Rise of Cynicism in Dental

cupation with the legal implications of health-care-giving. The questions of "underidentification," "overidentification," and "coincidence" of law and ethics, then, cannot be seen as too theoretical or trivial for further attention and action. Professor Dickens' paper reminds us that to deprive medicine of law, ethics, or both through inattention or faulty reasoning is nothing less than self-imposed impoverishment for us all.

II. INFORMATION FOR CONSENT: JUDGING THE DOCTOR'S JUDGMENT

In speaking of "The Standard, Functions, and Effect of Informed Consent," Professor Dickens defends the legal view of the ethical physician as one who, knowing the patient as individual and partner in the medical enterprise, acts to protect the patient's "human status." His summary comes in the comment, "The law not only requires ethical practice, but rewards it."

Ethics depends on private decision, professionally-applied; law depends on judicial determination of the "public standard" against which "private decision, professionally-applied" is to be measured. Rigorously pursued, given the variety of physicians and patients, these two standards allow for a wide spectrum of "adequate" or "reasonable" knowledge as prerequisite to the formation of informed consent. Each level within the spectrum, although absolutely different, is identical as part of it.

To insist in this situation that the legal is the ethical is to place the physician in an "impossible situation." In acting ethically, physicians know of their correctness as they judge. Legally, however, they can know this only after their acts are performed and judged. In retrospect, the ethical may be illegal. This Kafkaian situation has led some physicians to opt for independent (extralegal) physician-judgment as the method for guaranteeing "adequate" and "reasonable" knowledge in the informed consent process.[8]

Since judicial decisions are to some degree relative,[9] the ethical basis (seen as private, not to say prejudiced) must meet an apparently arbitrary standard. Relativity is compounded, so to speak. Further, in

---

School," *J. Health Social Behav.* **12**, 290–299 (1971).

[8]E.G. Laforet, "The Fiction of Informed Consent," *J. Amer. Med. Assn.* **235**, 1579–1585 (1976).

[9]R. Dworkin, "Symposium: Philosophy of Law—Judicial Discretion," *J. Phil.* **60**, 624–638 (1963).

insisting on this, the truly "ethical" may not be done, as judicial decisions are sometimes alleged to demonstrate. The possibility of replacing ethical judgment with reference to case law is also strengthened: in the teaching situation, development of an ethical standard (the basis of law in Professor Dickens' paper) may well mean "learning what the law will bear."

Consent presupposes communication. In terms of "informed" medical consent, that communication involves "adequate," "reasonable" knowledge, i.e., information given and received in a dually-individual sense: this individual patient agrees to participate in an action with this knowledgeable individual physician. What does knowledge (knowing the individual) entail?

1. The physician must not "under" or "over" inform, and thus must be quite certain of diagnosis and prognosis (or at least express whatever uncertainty is felt).[10] The patient's "uniqueness" must be grasped, e.g., that individual's "world-view", work, and family situation. How much else is necessary for adequacy? The physician must estimate the patient's general ability to understand medical information, and assess that ability in the present situation, ensuring such twofold comprehension because "adequacy" is the mark of free consent. Somehow, the information must be "tailored" with respect to its likely "emotional impact" so that physician-knowledge is accepted or rejected knowledgeably and voluntarily.

How will the physician gain such knowledge? Rejecting a retreat to Aesculapian intuition, will the medical curriculum prepare the candidate physician for this?[11] Is the technique for this knowing to be taught as theory, reinforced by exposure to suitable role models during undergraduate clinical practice, and assessed as "internalized" by certain measured criteria prior to conferral of the M.D.? If the ethical standard is to be the model enforced, how is the concept and practice of "doing what is best" to be inserted into the curriculum, and if it can be, will it be acceptable in the courtroom? Without some clear assurance that the required "knowing process" is possible, and that tangible results can be expected and will somehow be "uniform," the insistence on observance of an ethical standard (yet to be defined and quantified) seems unreasonable.

The same might be said of ensuring physicians' continuing discharge of this ethical (legal) responsibility. Patient and peer pressure

---

[10]S. Gorovitz and A. MacIntyre, "Toward a Theory of Medical Fallibility," *Hastings Center Report* 5 (6), 13–23 (1975).

[11]I am indebted to a remark of Professor J.M. Cameron here.

will be the main source for this, but ultimately, the judicial decision (to which the others may be reducible) will determine the matter. Again, legal insistence on an undefined ethical stance is somehow untenable.

2. Patients must also be under some ethical, not to say legal, compulsion to become knowledgeable in the course of this important process of communication that necessarily precedes the realization of "informed consent." The patient must know the physician's competence in diagnosis (by experience or reputation), and must be able to distinguish between the physician's personal and professional judgment. He or she must know when personal understanding is lacking, and persuade the physician to provide further information. At the same time, the patient must judge his or her emotional state so as to do freely and knowledgeably what is best. Is this possible? How will such a process be learned, much less guaranteed in practice? Although the form of consent may be present, its substance will nonetheless be lacking unless such a duality of communication exists.

The common good requires the ultimacy of law in society, and law is often compelled to "set up standards we cannot reach ourselves ... lay down rules we could not ourselves satisfy."[12] The common good also dictates the need for a balance of such standards with shared patient and physician initiative and responsibility. The difficulty of realizing "adequate," "reasonable" understanding in communicating knowledge between "individuals" is substantial. If the legal requirement of the ethical ideal is to remain, there must be some attempt to define and "quantify" the dual responsibilities involved. Protection of the "personal" may well depend upon the use of "impersonal" mechanisms.

III. SURROGATE CONSENT: PROCESS AND SUBSTANCE

Recent literature attests to ethicists' interest in the question of consent for the child research subject who will not personally benefit from the experimental process.[13] In his discussion, Professor Dickens notes the legal approach to this question, but makes no detailed comment about

---

[12] *The Queen* vs *Dudley and Stephens, op. cit.,* p. 288.
[13] Cf. R. McCormick, "Proxy Consent," *Perspect. Biol. Med.* **18**, 2–20 (1974); "Research on Children," *Hastings Center Report,* **6** (6), 41–45 (1976); P. Ramsey, "Children as Research Subjects," *Hastings Center Report* **6** (4), 21–30 (1976).

it.[14] He remarks that, legally, surrogate consent for the child is a "matter of process as well as of substance."

In view of what has been said in the essay, or perhaps more correctly, in view of what has been omitted, one must ask:

1. Will Professor Dickens not set further "guidelines" for consent in the nontherapeutic research situation in general, even as he has for the physician–patient relationship?

2. What does "process as well as substance" mean when applied to surrogate consent for the child in the nontherapeutic situation?

Professor Dickens' discussion of the "legalities" of research in general includes comments on the place of self-awareness of activity on the part of the physician, the need for information about "what is known . . . [and] not known" on the side of the subject–patient, the requirement that subject–patient 'enthusiasm' be held in check by the physician. In the context of this article, one might assume that the standard regarding information for consent is almost the same, whether research or therapy is contemplated.

In another essay, however, Professor Dickens develops the notion of a distinct difference of intent between the physician–therapist and the physician–researcher, as well as a clear difference between the ethical and legal principles involved in research and therapy.[15] Some difference in the nature of consent might then be assumed, and Professor Dickens remarks that "the critical question concerns the aspect of the research project to which (the subject's) consent must be directed."[16] The researcher's level of knowledge is a pivotal point here,[17] as is concern that "no detail of conceivable risk should be omitted from the information given to (the research subject) in seeking his consent."[18] Finally, Professor Dickens argues (contrary to earlier parts of his present essay) that "the law may be satisfied by practices violating ethical principles, in human experimentation as in much else".[19]

In brief, the present essay is not to be read without reference to his other writings. Professor Dickens apparently has espoused two differing views of consent in the research and therapeutic milieu in as

---

[14]Perhaps not his main point, the issue is implicit in the title of the essay. Professor Dickens' expertise in the area of experimentation is well-recognized. It would be helpful to the reader of this essay to have access to his articles on the subject.

[15]B.M. Dickens, "Information for Consent in Human Experimentation," *University of Toronto Law J.* **24**, 382, 385 (1974).

[16]*ibid.*, p. 393.

[17]*ibid.*, p. 396.

[18]*ibid.*, p. 397.

[19]*ibid.*, p. 387.

many papers. One can rightly ask: why and for what reason Has his position changed? Is there or is there not a difference between the consents involved? Or is this change a sign of the ambivalence earlier noted when the law was identified as "enforcing the ethical"? Clarification of the point is required if one is to appreciate the meaning of legally informed consent in both research and therapy.

"Consent for another" is contradictory on utterance of the phrase. In the therapeutic setting, the fiction prevails that this is not only possible, but desirable if the patient cannot personally consent. The argument rests on the need of the patient for care as superseding the normal requirement of personal agreement: in this case, the responsibility for "consent" must rest with a "disinterested" person, a "personally concerned" individual. Although such third party agreement may be the "process" of which Professor Dickens speaks, is it the "substance" (defined earlier as recognition of "individual autonomy")? Although perhaps necessary in certain circumstances, how can such surrogate "consent" be identified substantially with the personal consent given for therapy in less extreme conditions?

A more serious question surfaces in consideration of the "substance" of surrogate consent for the child subject in the experimentation process not directed to that child's benefit. If "substance" cannot be conceded in the earlier case, is it more acceptable in this? Going further, can one even argue that "due" "process" has been observed in such a situation? Once more, Professor Dickens has not discussed the problem directly here, but by inference, one assumes that both "process and substance" can be satisfied legally by surrogate consent for the child involved in nontherapeutic research.

On recourse to the other essay of Professor Dickens, however, it is clear that this point must be reconsidered. Arguing that "sufficient emergency (could) justify risking the lives of others in an act of social defence both with or without their consent,"[20] Professor Dickens also comments that "It may be doubted that the law has yet advanced its procedures to regulate which of our fellow-creatures may become the subject of experimentation to save the residue."[21] Going beyond surrogate consent to consider social emergency (though with reservations), Professor Dickens exceeds even this criterion when he notes elsewhere: "One must set against the individual interest in not using children for experimentation the cost to children and to the community of which they are a part of the

---

[20]*ibid.*, p. 409.
[21]*ibid.*, p. 410.

experimentation not being undertaken."[22] "Sociality" thus becomes the norm: surrogate consent may or may not accompany it. Will there be any limit to this escalation?

The ethical nature of surrogate consent for nontherapeutic research involving children is still debated by physicians[23] as well as the ethicists. Professor Dickens' provocative comments in the essays considered are a positive contribution to that discussion. However, they require clarification and expansion so that inconsistency may be avoided. By implicitly raising this last question, Professor Dickens makes resolution of the original unasked question even more urgent: what is the basis for the distinction/relation between law and ethics?

---

[22]B.M. Dickens, "The Use of Children in Medical Experimentation," *Medico-Legal J.* **43**, 171 (1975).

[23]"Valid Parental Consent," *The Lancet,* 1 for 1977 (June 25, 1977), p. 1347.

Involuntary Commitment of the Mentally ILL: Some Moral Issues

Dan W. Brock

Department of Philosophy, Brown University, Providence, Rhode Island

In considering some of the distinctly moral issues raised by the treatment of the mentally ill, perhaps the most striking fact is the frequency with which coercion in one form or another is used on the patient and treatment is as a result involuntary. In this article I shall discuss several of the moral issues that this use of coercion creates, with particular emphasis on the justification of involuntary civil commitment. My aim will be to outline some of the underlying moral principles and distinctions on the basis of which such issues can be profitably discussed and clarified. Since involuntary commitment of the mentally ill has frequently come under attack, I will be particularly interested in whether certain of the common arguments against it are sound. Although laws governing civil commitment provide an obvious point of departure for much discussion in this area, I will be concerned here with the moral positions that ought to form the basis of legal policy. Decisions regarding the involuntary commitment of persons believed mentally ill are commonly agonizing and troubling, with deep emotional involvement, conflicts, and impacts on all those a party to them. Because of the highly charged atmosphere in which these decisions must often be made, it is all the more important that the relevant moral issues and principles be clearly delineated.

A helpful method for shedding some light on the moral justifica-

---

*An earlier, longer version of this article has appeared in *Philosophy and Medicine,* vol. **5,** *Mental Illness: Law and Public Policy,* Baruch Brody and H. Tristram Engelhardt, Jr., eds., Dordrecht, Holland, Reidel, 1978.

tion for involuntary civil commitment of the mentally ill on grounds of dangerousness, perhaps the most common ground for involuntary commitment, is to contrast civil commitment with penal commitment, and the moral justifications commonly offered for it. Let me here briefly enumerate the usual justifications offered for criminal punishment in order to see whether they also apply to involuntary civil commitment, and when they do not, what differential circumstances in the involuntary civil commitment of the mentally ill are required to reflect this.

1. *Deterrence.* The threat of punishment for performing a criminal act is designed to deter persons who might otherwise be tempted to perform the act from so doing.

2. *Protection.* Criminal punishment, at least in the form of imprisonment, also has the purpose of protecting society, specifically other members of society, from dangerous persons and the harms they create. In criminal law, protection can only be provided against persons convicted of past criminal acts; dangerousness, insofar as it justifies protection by imprisonment, is not a predictive category in the criminal law, as it is in civil commitment laws. The purpose of protection served by criminal punishment is importantly limited by other principles of the criminal law, since prisoners cannot be held beyond the terms of their sentence, however dangerous they may still be, and cannot be incapacitated in order to make them nondangerous.[1]

3. *Rehabilitation.* Rehabilitation or reform of the convicted criminal is designed to change an individual's character, motivations, opportunities, skills, and so forth, so that the law will not be violated again upon the prisoner's release. The forms it may take are limited by the rights an imprisoned criminal retains. In general, rehabilitation is designed to *improve* prisoners and their personal situations, not simply to render them nondangerous, which latter effect could of course be achieved through various kinds of massive incapacitation, the use of which would outrage most of us.

All of these first three purposes of, or justifications for, criminal punishment fit a general utilitarian or preventive model of the criminal law as a method of reducing the incidence of certain unwanted actions.[2]

In this general preventive account of penal commitment, certain striking differences in the penal and involuntary civil commitment

---

[1]The use of prison sentences for a range of years, such as 1–10 years, coupled with the present U.S. parole system, is an important qualification on this, though fixed-term sentences seem currently to be regaining favor.

[2]Readers unfamiliar with the literature and issues concerning utilitarianism may consult any ethics textbook, or my paper, "Recent Work in Utilitarianism," *Amer. Phil. Quart.* **10**, 241–276 (1973).

processes seem especially troubling. In particular, two arguments against involuntary commitment of the dangerous mentally ill are commonly made. First, it is a commonplace of Anglo-American jurisprudence that it is better for ten guilty persons to go unpunished than for one innocent person to be punished. Various rules of evidence and procedure within the criminal justice system are designed to insure that when we fail to convict all and only the guilty, as we inevitably will, we fail to convict the guilty far more often than we mistakenly convict the innocent. Available evidence suggests, however, that in the civil commitment process our toleration of error is exactly reversed. Studies indicate that when psychiatrists find mentally ill persons to be dangerous, or likely to cause substantial harm to themselves or others, they tend greatly to overpredict dangerousness.[3] For each failure correctly to predict as dangerous an individual who subsequently commits an act causing serious personal harm to others, numerous other persons will be judged dangerous who do not in fact turn out to be such; in statistical terms, the number of false positives is substantially greater than the number of false negatives. Why do we prefer to let the guilty go free rather than wrongly convict the innocent under the criminal law, although preferring in dealing with the mentally ill to commit involuntarily the harmless rather than let a dangerous person go free? Can there be any justification for this strikingly increased willingness wrongfully to confine the mentally ill as opposed to the convicted criminal?

The second striking difference leads to a more wholesale attack on the involuntary commitment of the mentally ill for dangerousness. With very limited exceptions, preventive detention—the detaining of persons who either have committed no offense or have not yet been convicted of committing an offense—is not permitted in the criminal justice system.[4] But *all* involuntary commitment based on a prediction that a person is dangerous, that is, will at some point in the future commit an act causing serious personal harm or harm to others, is preventive detention. Even in states where such predictions must be based in part on past harmful acts, involuntary commitment is still preventive detention since the commitment is not for the past acts, but rather the past acts are intended only to serve as strong evidence for the likelihood of future harmful acts. But if preventive detention is

---

[3]Cf. A. Dershowitz, "Psychiatry in the Legal Process: A Knife that Cuts Both Ways," *Trial* **4**, 32 (1968).

[4]Persons often do spend significant periods of time in detention before trial, but except in limited circumstances, bail must be set for them, the payment of which will secure their release until trial.

abhorrent in the criminal law process, why do we allow it as a common-place, as the *standard procedure,* in the involuntary commitment of the mentally ill on grounds of dangerousness? Can there be any justification for denying the mentally ill this important protection afforded criminals?

There is a fourth justification of criminal punishment that provides a partial explanation of these differences, and sheds light in turn on the justification of the involuntary commitment of the mentally ill—a justification from justice or fairness. The legal system can be seen, at least ideally or when it is just, as setting up a fair or just distribution among citizens of benefits and burdens, of rights and duties or responsibilities.[5] In particular, the criminal law requires everyone to forbear from certain specified acts that cause harm to others, and results, so far as the law is obeyed, in everyone being free from having those harms caused to themselves. The criminal then acts unfairly in receiving the benefits of others' forbearance from harm, even while failing to forbear from harming others. Criminal punishment, therefore, is required in order to help restore the just balance of benefits and burdens that the criminal law ought to establish, and to prevent the criminal from profiting from wrongdoing. This provides at least part of the basis for the principle that violators of the criminal law *deserve* punishment, one form of which may be loss of liberty by imprisonment, a principle that the deterrence, protection, and rehabilitation justifications fail adequately to account for. Punishment is *required* by justice or fairness, and need not simply be revenge in the absence of any of the three utilitarian sorts of justifications.

There is a second aspect of the justice or fairness justification. The criminal law can be understood as saying to each of us, in effect, refrain from prohibited acts such as homicide, or suffer punishment.[6] The excuses that in the criminal law remove a law-breaker's liability to punishment—roughly, coercion, inability to conform to the law, and nonculpable ignorance of the nature of one's action—are designed to insure that the violation of the law was both intentional and voluntary. They are designed to prevent violation of the moral principle that persons should not be punished who could not reasonably have helped

---

[5]This view of justice is elaborated at great length and very powerfully in John Rawls, *A Theory of Justice,* Cambridge, Mass., Harvard University Press, 1971; it is specifically applied to criminal punishment in, among other places, Jeffrie Murphy, "Three Mistakes About Retributivism," *Analysis* **31,** 166–169 (1970–71).

[6]This conception of the law as a choosing system has been made well known by H. L. A. Hart in his *The Concept of Law,* Oxford, Oxford University Press, 1961, and his *Punishment and Responsibility,* Oxford, Oxford University Press, 1968.

doing what they did. This conception of the law as a choosing system helps explain the basis for the principle that *only* the guilty should be punished, that only lawbreakers *deserve* punishment. The law, in effect, says abstain from proscribed acts or be punished, and it implicitly promises that if one does refrain from performing proscribed acts, one will be free from punishment. Therefore, it would be a serious unfairness or injustice to punish those not found to have been lawbreakers.

The two striking differences pointed out earlier in our treatment of criminals and the mentally ill should now begin to make at least some sense. We should be able to say that the criminal has acted freely, knowingly, and unfairly in performing an expressly and publicly forbidden act, and for most of the criminal law, performed an act that is in itself morally wrong (homicide, assault, fraud, etc.) apart from the charge of unfairness; there is as a result an expressive function to a finding of guilt in the institutional process of the criminal justice system, a public expression of disapproval and condemnation of the action and character of that individual.[7] It is this condemnation, with its affect on a person's reputation, as well as the loss of liberty present in both civil and penal commitment, that the bias against punishing the innocent is designed to avoid. In the case of an innocent person, such condemnation is a false charge. Involuntary civil commitment on grounds of dangerousness, on the other hand, is not punishment for an action for which the mentally ill person can fairly be held responsible, and so the public condemnation essential to punishment is absent. This provides sound reason for at least *some* greater concern to avoid punishing the innocent, in order to avoid such a false condemnation, in the penal than in the civil process.

The other difference, the use of preventive detention in civil commitment, should now make a bit more sense as well. Given the "forbear or be punished" model of criminal law, only those who have been found to be violators can fairly be punished. Again, it is this unfairness, as well as the loss of liberty, that is objectionable in preventive detention; preventive detention is incompatible with this model of the law. The law is designed to allow persons to control and predict when they will encounter sanctions such as fines and imprisonment by making their imposition follow only upon choices known beforehand to have that consequence. Preventive detention breaks this link between choice and sanction, and so destroys our ability to predict and control when we will be subject to such sanctions. There is no analogous requirement of

---

[7]Cf. Joel Feinberg, *Doing and Deserving,* Princeton, Princeton University Press, 1970, chapter 5.

accountable choice in the civil commitment process for preventive detention to be incompatible with. The conditions of accountable choice are just what the mentally ill are considered incapable of satisfying. Nor is involuntary civil commitment for dangerousness, when no harmful act has yet been performed, punishment of the innocent in the same sense that this latter notion possesses in the criminal process. Since the condemnation inherent in the penal process is lacking, there is no claim made that the innocent person is guilty; there is no punitive element, no punishment, and therefore no attribution of guilt or innocence.

Although preventive detention would thus have especially objectionable features if used in the criminal law, it is still not clear why we are justified in abandoning, when we deal with the mentally ill, the restriction observed in criminal law whereby only those who have already committed a criminal act may have their liberty taken away. Why is a preventive approach to harmful acts by means of involuntary detention justified with the mentally ill when it is not considered such for other citizens?

We can use the law to reduce the incidence of specified unwanted acts because adult humans in general have the capacity to guide their behavior in accordance with promulgated social rules. It is, roughly, the capacities to form purposes, weigh alternatives according to how they fulfill those purposes, and act on the result of that decision-making process that renders it possible to redirect human behavior with promulgated social rules, specifically, legal rules carrying sanctions. Our use of this method of controlling criminal behavior, which may at least in some kinds of cases be less effective than a predictive/preventive approach, reflects a number of factors, among which are: first, the important value ascribed to individual autonomy and choice—to persons controlling, being responsible for, and held accountable for their own behavior; second, the importance given to the individual's rights of liberty and privacy, both in our moral thinking and in our constitutional tradition; third, the great potential for error and abuse present in a system that would allow any group of persons, whether government authorities or physicians, to deny other persons their liberty on a prediction that they might at some point commit criminal acts.

The dangerous mentally ill are treated differently, I believe, at least to the extent that the difference purports to be justified, because they are not considered capable of guiding their behavior by promulgated social rules in this way. As a result, they are not subject to the "do it or else" criminal law, but rather to a cost/benefit calculus of protection and prevention. To put the point dramatically, we lock them up as we would a rabid animal because their behavior cannot be guided and controlled by the more desirable means of laws. Let me emphasize that my point is *not* that the mentally ill are like animals, but rather that when we abandon the criminal law system appropriate for *persons,* in favor of the preventive model, we treat dangerousness in the mentally ill, and so the mentally ill themselves, as we treat dangerous animals and things. If this is the underlying assumption, then a careful sorting out of different forms of mental illness, and of mentally ill persons, is required with regard to the effects of those illnesses on a person's capacity to understand and follow legal rules. In fact, much that is now classified as mental illness seems not substantially to impair that capacity, at least below some reasonable minimum level. Just when these capacities are seriously impaired is a difficult empirical question that needs investigation—cases in which the thought and behavioral processes of individuals are so disordered that they are incapable of understanding rules, weighing evidence, appreciating the consequences of their action, and controlling their behavior, are the clearest examples of what I have in mind. Without such an impaired capacity to be guided by legal rules, dangerousness in the mentally ill ought to be dealt with just as it is in the nonmentally ill—punished through the criminal process after illegal acts occur, but not used as a ground for loss of liberty before such acts occur.

If most current involuntary commitment processes are indefensible in important respects, as I believe they are, it is crucial that we correctly understand why this is so. I have defended two of the important differences distinguishing civil commitment from penal commitment—the departure from the strong bias toward not convicting the innocent or harmless, and the use of a preventive approach, preventive detention, with *some* mentally ill persons. These are common lines of attack against involuntary commitment, but they are not decisive. If a preventive approach to harmful acts is justified for some, though by no means all, mentally ill persons, do we prevent greater harms than we create by involuntary commitment of persons judged mentally ill and dangerous? If we do, then we have found plausible justification, and have cleared away some common and seemingly plausible objections for doing so. The difficulty is that there is only rarely, if ever, sound reason to believe that we produce a favorable balance of harms over benefits with involuntary commitment. The problem lies not in the absence of a plausible moral basis for involuntary commitment, but in a failure to satisfy the empirical requirements of that basis.

The problem lies in the predictability, or lack thereof, of dangerousness. Dangerousness is rarely, if ever, predictable in the mentally ill with sufficient accuracy to justify prior commitment. Several points need to be made in this context. The first is that the category of dangerousness must be narrowed down and made precise. The complete loss of one's liberty involved in involuntary commitment is a very substantial harm. Acts to be prevented under the category of dangerousness must at least involve harms of the same dimension. Persons who exhibit bizarre behavior, such as loud and incoherent talking to strangers in public places, are certainly often a nuisance and their behavior makes most persons quite uncomfortable. Nevertheless, I believe it is clear that no harm that they cause comes close in magnitude to the harm of their loss of liberty if involuntarily committed. Offensive behavior, such as indecent exposure, also should be excluded from the category of dangerous behavior—being offended, although perhaps a harm, is again not a harm of sufficient magnitude to justify involuntary commitment; the same holds for behavior that is disruptive. The relevant acts under the category of dangerousness should be homicide, rape, and other serious acts of violence that cause substantial physical or psychical injury.

The next point is a simple, but crucial, statistical one.[8] When the category is narrowed in this way, dangerous acts are of quite infrequent occurrence, and infrequent acts create special problems of prediction. Even an extremely accurate test or predictive procedure for identifying people as dangerous or nondangerous will produce far more false predictions that a person is dangerous (in statistical terms, false positives) than correct predictions that a person is dangerous (true positives). For example, suppose, not unreasonably, that only 1 out of every 1000 persons will commit a serious act of violence of the required sort. And suppose we have a test or predictive procedure that is accurate 99% of the time in sorting out persons as dangerous or nondangerous; that is, it correctly identifies the persons in fact violent 99% of the time, and correctly identifies the persons in fact harmless 99% of the time. This is vastly more accurate than any predictive methods we now possess. Suppose 1,000,000 persons are tested. This test will correctly identify 990 of the 1000 dangerous persons, and by involuntary commitment we could prevent the future harms they would cause if left at liberty. But the test will also identify 9990 other persons as dangerous who in fact are harmless. To prevent the harms that the 990 will cause, we must

[8]Cf. P.E. Meehl and A. Rosen, "Antecedent Probability and the Efficiency of Psychometric Signs, Patterns, and Cutting Scores," *Psych. Bull.* **52**, 194–216 (1955). A. Rosen, "Detection of Suicidal Patients: An Example of Some Limitations in the Prediction of Infrequent Events," *J. Consult. Psych.* **18**, 397–403 (1954).

inflict on the other 9990 harmless persons the very serious harm of the loss of their liberty.[9] I do not see how any utilitarian or cost/benefit calculus could support involuntary commitment based on such procedures. Of course, if the mentally ill were far more dangerous than the public at large, so that among them the incidence of violent acts was far greater, then the ratio of those in fact harmless to those in fact dangerous, who are identified as dangerous, would be reduced. However, available evidence indicates that the mentally ill are overall not significantly more dangerous than the general public.[10] It may, however, be possible to identify narrower categories of mental illness, for example certain forms of paranoia, associated with a considerably higher frequency of dangerous acts of the required sort, though the considerable imprecision and vagueness in many categories of mental illness is a hindrance to establishing such associations. Although sufficient predictive accuracy to justify involuntary commitment seems at present not to exist for many, if any, narrower categories of mental illness, it is only such narrowed categories that hold any significant future prospect for adequate prediction.

Since the harm of the loss of liberty to the person in fact not dangerous is a very serious one indeed, the burden is on those who would use any predictive technique for dangerousness to establish that it possesses the necessary accuracy before it is employed in involuntary commitment proceedings. The point holds equally for any clinical method of establishing dangerousness. Just what the necessary degree of accuracy should be is difficult to say since the actual calculations required are more complex than I have suggested, and must include such factors as whether and in what ways a dangerous mentally ill person can be made nondangerous, what period of involuntary commitment is likely to be required, and so forth. However, I believe the conclusion stands that present or foreseeable predictive techniques for dangerousness are rarely, if ever, sufficiently accurate to justify their use to deny the mentally ill their liberty, certainly for anything beyond extremely short-term emergency hospitalization. Predictions of dangerousness could more easily justify involuntary treatment when treatment

---

[9]This example is taken with slight alteration, from J.M. Livermore, C.P. Malmquist, and P.E. Meehl, "On the Justifications for Civil Commitment," *Univ. Pennsylvania Law Rev.* **117**, 75–96 (November, 1968).

[10]Cf. Giovanni and Gurel, "Socially Disruptive Behavior of Ex-Mental Patients," *Arch. Gen. Psychat.* **17**, 397–398 (1972). Rappeport and Lassen, "Dangerousness—Arrest Rate Comparisons of Discharged Patients and the General Population," *Amer. J. Psychiat.* **121**, 776 (1965).

does not require confinement of the patient; this is simply an instance
of the general principle that, other things equal, the least restrictive
treatment alternative is always to be preferred.

Until now I have been concerned with involuntary commitment of
mentally ill persons on grounds of dangerousness. Much involuntary
civil commitment, however, is carried out "for the patients own good."
This may involve the prevention of self-inflicted harm, but it need not.
Categories such as "in need of care or treatment," "such that one will
benefit from treatment," "in need of treatment and lacking sufficient
capacity to recognize that need" are common in state commitment
statutes and suggest justification solely on grounds of benefits to be
provided to the patient committed. I want to say a little about this sort
of involuntary commitment by discussing briefly the issue of paternalism. I shall suggest what such a paternalistic principle would be like in
a moral theory that takes a right to liberty of the person or privacy, as
one basic moral right. Roughly, paternalism is action performed by one
person, A, for the good of another person, B, that violates B's moral
rights, and that is justified independently of the past or present consent
of B. When is it morally justified? One way of posing this question is
to ask when we would want others to so act on our behalf, when we
would consider them justified in so doing. Moral rights are intimately
connected with the notion of autonomy or self-determination and the
placing of important value on being autonomous. Our concept of autonomy applied to persons depends on our conception of persons as able
to form purposes, to weigh alternative actions according to how they
fulfill these purposes, and to act on the result of this deliberative process. And being autonomous in this sense is something many people
value for its own sake, as part of an ideal of human excellence, and not
simply as instrumental to human happiness or some other goal.

We will want others to act paternalistically toward us, and therefore in ways that violate our moral rights, only when our capacities to
form purposes, weigh alternatives, and to act on the result of that
deliberation are defective. Otherwise, we will want the protection of
uncoerced choice and action that moral rights provide. As one philosopher has put it,[11] we will want others to act paternalistically toward us
only when we are subject to an evident weakness or defect of will or
reason. Since such failures of will or reason are a necessary, but not a
sufficient, condition for the justified use of paternalism, no belief that
persons will be made better off by our action will ever be sufficient in
itself to justify our acting paternalistically toward them.

---

[11]Rawls, *op. cit.*, p. 249.

By a failure of will I mean the inability to act as we believe we ought, all things considered, to act, that inability arising not from our inability successfully to perform the action once we have undertaken to do so, but from a failure to be sufficiently motivated to undertake the action. Cases of extremely strong temptation, and physical or psychological addiction, are examples. These will generally be cases in which, while subjectively experiencing the situation, the patient feels unable to understand and/or control his or her behavior, and does not identify with the behavior in the sense of wanting to be motivated in that way. Our reasoning in such cases should be based on our knowledge of the person in question and on an estimation of what that individual would consent to were it not for the specific failure of will affecting his or her present behavior. An adequate principle of paternalism will in general require evidence that, except for a defect of reason or will, persons would waive the rights that our action violates, and would choose to have us act in ways that violate that right. I shall not elaborate here on specific cases largely because I am inclined to believe that usually, when there is sufficient evidence of failure of will to justify the very substantial invasion of liberty that involuntary civil commitment represents, it is possible to obtain the patient's consent to the hospitalization, certainly for anything beyond short-term emergency hospitalization; when consent is obtained, we do not have paternalism.

There are several sorts of defects of reason. Consider first defects in knowledge. For example, imagine an individual is under a delusion of personal invincibility and is about to step into the path of a speeding car believing that no harm will result from so doing. Certainly, assuming we have no independent reason to believe that this individual wants and has good reason to suffer harm, but only mistakenly believes that no harm will occur, we would not hesitate to use coercion to push so deceived a person from the path of the car. We would be justified in this because the person will not obtain what is wanted in acting on such false beliefs. With a normal person, we could provide evidence on the basis of which the person would change the mistaken belief, and would in turn give up the desire to perform the harmful action—no coercion would normally be necessary. The defect is not simply holding false beliefs—all of us do that—but rather being unable to conform one's beliefs to available evidence. The importance of this sort of defect of reason for mental illness should be obvious—much psychosis including, most obviously, severe depression, mania, and paranoia, result in a person adopting and steadfastly maintaining beliefs whose falsity should be readily apparent, beliefs that often are in turn causally responsible for one's harming oneself, or refusing clearly beneficial treatment.

The first difficulty, both theoretical and practical, in this category arises from the necessity of distinguishing among failures of knowledge, defects in one's reasoning concerning factual or empirical matters, and differences in values. If what a person wants and the kinds of activities that are valued can only be *different* from what others want or value, but cannot be valid or invalid, or correct or mistaken, then it seems that the person with unusual values or desires has no defect or deficiency of reason that might reasonably justify another's paternalistic action. I believe this view that desires and values only differ, but are never subject in themselves to rational evaluation, is mistaken. Most highly unusual desires are no more than that—highly unusual, perhaps even peculiar, but not irrational; for example, tastes for strange foods or driving motorcycles across canyons.

However, all desires or values are not simply different one from another. The objects of some desires are, in themselves, irrational. For example, to suffer great pain or to be seriously injured is intrinsically bad for a person; these are intrinsic evils.[12] That is not to say that it may not be, all things considered, rational to desire one of these, for example, if one's life can only be saved at the cost of serious injury. But to desire to be seriously injured for its own sake and for no further reason is irrational.[13] Possession of an irrational desire of this sort is, then, another kind of defect of reason. I would suggest that paternalistic action can be justified to prevent a person from acting on an irrational desire when the likely harm or evil, or loss of benefit, to him from doing so is very substantial.

Not only can the objects of some desires be irrational, but some weightings of gains and benefits, and of conflicting desires, can be as well. There is much room for reasonable difference about the relative

---

[12]The examples of intrinsic evils are from Bernard Gert, *The Moral Rules,* Harper and Row, New York, 1970.

[13]As I now see it, an account of the nature of irrational desires would appeal to the notion of persons as purposive beings, and to certain objects or states of affairs as useful to the promotion of the range of purposes characteristic of persons. (It should be noted that some contingent truths of human psychology will be needed here.) Very roughly, it will be irrational, other things equal, not to desire such things, or to desire their frustration. This, of course, is *far* too simple, but the present paper is not the place to take up this quite difficult question; I hope to say more about the nature of rational and irrational desires on another occasion. Among useful discussions by philosophers of this question are: John Rawls, *op. cit.,* ch. 7.; Phillipa Foot, "Moral Beliefs," *Proc. Aristotelian Soc.* **59**, 83–104 (1958–1959); Richard Brandt, "Rational Desires," *Proc. Amer. Phil. Assn.* **43**, 43–64 (1969–1970); Bernard Gert, *op. cit.*

importance of a particular harm or benefit, and in turn about the overall desirability of various choices, actions, and activities. Such differences provide no justification for paternalistic action, for imposing one's own conception of the good on another. But other weightings of the relative importance of desirable or undesirable aspects of an activity or action are such that they do admit of criticism from our notion of rationality. Consider, for example, a person whose life could be saved only by an injection, who wanted very much to live, but who refused the injection because of fear of needles. Weighing the discomfort from the injection as more important than the loss of one's life, without the existence of some further special background circumstances to change the case, is irrational. This provides us with another form of defect of reason. The very difficult distinctions that must be made here, which difficulty itself creates great potential for abuse in this category of paternalism, are between irrational desires, and irrational weightings of harms and benefits, as opposed to desires or weightings that are simply extremely unusual. When paternalistic action is necessary to prevent action on irrational desires that will result in serious and not easily reversible harm or loss of benefits, and when there is good reason to believe that if and when a person is no longer subject to this defect of reason, that person will come to support our intervention, paternalistic action can be justified.[14]

What is common to all defects of reason and will is a deficiency in the ordinary human capacity to form purposes, to weigh alternative actions according to our conception of our interests, and to choose to act accordingly. Paternalism is never justified in the absence of this deficiency. It is because mental illness can so seriously deform a person's reason, will, and action from what they otherwise would have been that paternalistic justification of involuntary hospitalization and treatment is possible and important.

I want to add that given the difficult and problematic nature of the distinctions I have used, and the resulting very substantial potential for abuse in the employment of paternalism on these grounds, we might

---

[14]More than one psychiatrist has maintained to me in discussion of this issue that desires to possess, for their own sake, things intrinsically bad for a person, what might be called pure irrational desires, do not exist. Their view is that seemingly pure irrational desires of this sort are always based on false beliefs (and, on some views, in many cases unconscious beliefs) which if true would render the desire not irrational. Whether this is correct is, or at least should be, an empirical matter; even if correct, it does not show that irrational desires are not important in the area of mental illness, but only that pure irrational desires are not. In any case, I believe the category of pure irrational desires remains important for an understanding of the issues in this area.

reasonably decide as a matter of public and legal policy not to grant
anyone the right to act paternalistically, or to greatly restrict any such
right, although nevertheless accepting the moral position I have proposed.

Our brief examination of involuntary commitment of the mentally
ill on grounds of dangerousness to others, and on paternalistic grounds,
has yielded one rather unexpected general conclusion that is worth
stressing. It has seemed to many commentators in this area that if
involuntary commitment of the mentally ill is justified at all, it is most
clearly justified when the person is judged to be dangerous to others,
whereas paternalistic justifications are commonly considered morally
suspect. Paternalistic commitment is not devoid of important prediction problems of its own, but at least in cases of the provision of
important benefits through treatment there may be a significant range
of cases in which those problems are considerably less severe than with
commitment for dangerousness. These cases of paternalistic commitment may therefore be, contrary to the common view, less morally
problematic than most commitment on grounds of dangerousness to
others.

Comments on Brock's "Involuntary Commitment of the Mentally ILL: Some Moral Issues"

Bruce L. Miller

Department of Philosophy, Michigan State University, East Lansing, Michigan

Professor Brock concludes his paper with the following claim:

> . . . cases of paternalistic commitment may therefore be, contrary to the common view, less morally problematic than commitment for dangerousness to others.

I will argue that Brock is wrong, and that the common view is wrong. I will defend the position that paternalistic commitment and commitment for dangerousness to others are equally problematic and for the same reasons. In the process I will accept much of Brock's analysis, some because I agree and some arguendo.

Whether commitment is paternalistic or for dangerousness to others, two similar judgments must be made: *1.* a judgment that the individual is mentally ill, and *2.* a judgment that because of such illness the individual is likely to perform a harmful act. The judgment that mental illness is present will not differ in the two cases. In the section on commitment for dangerousness to others, Brock discusses mental illness in terms of the capacities to form purposes, weigh alternatives, and act on the result of that decision process. Not all mental illnesses, he contends, are sufficient for commitment. In the section on paternalistic commitment, Brock discusses mental illness in terms of weakness of will and defects of reason. At the conclusion of the discussion of paternalism, Brock connects these two approaches when he says that what is common to all defects of reason and will is a deficiency in the ordinary human capacity to form purposes, to weigh alternatives, and to choose to act accordingly. It seems then that whether commitment is for danger to others or danger to self, the criteria for mental illness will be the same. Hence, the moral difference between them cannot be found in the judgments of mental illness.

In each of the two sorts of commitment, a judgment that as a result of mental illness the individual is likely to perform acts that are harmful must be made, either that the acts would be harmful to others or harmful to self. In discussing the former, Brock raises the familiar problem of the abundance of false positives in predictions of dangerousness. The moral problem in commitment for dangerousness lies not in the absence of a plausible moral basis for involuntary commitment, Brock argues, but in a failure to satisfy the empirical requirements of that basis. Brock does not argue, and I do not know of any reason to believe, that predictions of danger to self are less suspect on this ground than prediction of danger to others. Hence, the moral difference between them cannot be found in the judgments of dangerousness.

The difference that Brock sees between the two sorts of commitments is that in paternalistic commitment there will be cases in which important benefits are provided through treatment. Brock presumes that when we commit on paternalistic grounds we will benefit an individual in two ways: one will be to prevent the immediate harm to self by placing the individual in custody; the other will be to treat, and thereby benefit, the individual by release from custody and release from the disposition to self-harm. What Brock seems to imply here is that even though some individuals will be committed on paternalistic grounds who would not have harmed themselves, this cost may be cancelled, at least in part, by the benefits of treatment. Such benefits could go to the false positives as well as to the true positives, i.e., an individual who is predicted likely to inflict self-harm and yet would not do so can benefit from treatment just as an individual who is predicted to inflict self-harm and who would do so.

But why should the benefits of treatment not be taken into consideration when commitment is for dangerousness to others? There is as much reason to treat an individual who is a danger to others as an individual who is a danger to her- or himself. In both cases the individual will benefit by release from custody and there will be the benefit of reducing the likely incidence of harmful behavior. Brock presents no moral or empirical argument that this plausible view is incorrect. Thus, we have no reason to believe that paternalistic commitment is less morally problematic than commitment for dangerousness to others. In order to present my argument that they are equally problematic and problematic for the same reasons, I want to contrast civil commitment with penal commitment.

It is a common view that penal commitment for harm to others is less morally problematic than penal commitment for harm to self. It is also commonly believed that this view cannot be maintained on utilitarian grounds, as Mill seemed to think. For if we are operating with the utilitarian injunction to minimize harms, the question of whether the harm of commitment is outweighed by a resultant reduction in overall harm will be answered no differently for harms to others and harms to self; they will both count equally in the calculations. The argument against paternalistic interference by the criminal law flows from the proposition that liberty of action is valuable in itself, apart from any of its beneficial consequences. If X harms Y and X is punished by commitment, the harm to X and the interference with X's liberty of action is weighed against the harm to Y, the deterrence of future harm to others, and the violation of Y's right to liberty of action as well as the deterrence of future violations of right to liberty of action, because actions that are harmful to another usually limit liberty of action. But if X harms X, the consideration of violation of the right to liberty of action by the harmful action of X is not present, for an individual cannot violate his or her own rights. Hence, if the criminal law recognizes liberty of action, it should not interfere with conduct that is chosen, and harms no one except the actor.

What are the implications of the value of liberty of action when commitment is civil commitment rather than penal commitment? What this question comes to is the question of what are the implications of the fact that civil commitment is of individuals who are mentally ill rather than mentally competent, and who are likely to commit a harmful act rather than having committed a harmful act? Each of these differences provides a reason why some of the arguments for penal commitment are not available for civil commitment. First, since the conduct is not chosen (because of the actor's mental illness) the deterrence argument that the threat of commitment will reduce the likelihood of such conduct in the future cannot be made. As Brock puts it, the mentally ill are not subject to the "do it or else" criminal law. Second, because commitment is for harm likely to be done in the future and not for harm already done, it cannot be argued that the harm of commitment is justly deserved. The crucial factor in the justification of civil commitment is mental illness. It's role in justifying commitment is the same whether commitment is to prevent danger to others or danger to self. If the mental illness required for civil commitment is as

Brock and I urge, *viz.,* inability to choose or not choose the dangerous action, then there is no violation of the right to liberty of action. I will try to be persuasive on this claim by the use of examples.

Suppose that X has terminal cancer, has known it for some time, and is now approaching the last months of life. X wishes to avoid the pain and indignity of fighting to the last and therefore chooses to commit suicide. Y, knowing of X's decision, takes steps to prevent X from doing so. Now Y has surely interfered with X's liberty of action; Y may or may not be justified in doing so, but it clearly is an interference with what X has chosen to do.

Suppose that X is walking along a mountain trail and Y grabs X just as X stumbles over a rock and is about to fall to a certain death. In this case Y has not interfered with X's liberty of action, for X did not choose to die falling from the cliff.

But now suppose that X's family has just fallen off the cliff to their certain death and X feels responsible for their deaths. Full of grief and guilt, X attempts to jump from the cliff. Y, as in the above case, seizes X and prevents the death. Is this case one of Y interfering with X's liberty of action, although justifiably, or is it a case in which Y has not interfered with X's liberty of action because, as in the second case, X did not choose to die. If we take the view that there is mental illness in which an actor lacks the capacity to make choices, then we are committed to the further view that there can be cases, perhaps something like this one, in which preventing the conduct of another is not interfering with that person's liberty of action, and further that it would be mistaken to describe such cases as justified cases of interfering with the liberty of another.

The above examples are all cases of one person preventing another from self-inflicted harm. But I believe that the same point can be made with cases of one person harming another. Thus, if I can then consider myself armed with the claim that some cases of preventing the mentally ill from harming themselves or others is not an interference with their liberty of action, then I can move to my claim that there is no moral difference between civil commitment for dangerousness to others and paternalistic civil commitment.

It has already been argued that the problems of defining mental illness will be the same for commitment for danger to others and commitment for danger to self; it has also been argued that the problems of prediction are of the same magnitude whether the predictions are of danger to self or danger to others. Thus, on these two points, both sorts of commitment pose the same moral problems. Third, if it is not an interference with liberty of action to prevent someone from harming either self or another, because of the actor's mental illness, then that factor making paternalism more objectionable than protection of others when the actor is mentally competent is not present, and no claim can be made that one is more morally objectionable than another. Lastly, it was also argued above that there is as much moral reason to treat those dangerous to others as those dangerous to themselves. Thus, there appears to be no reason to believe that in civil commitment, unlike penal commitment, paternalism is more morally objectionable than protectionism.

In closing I would like briefly to mention two points that explain why it might be thought that paternalistic civil commitment is more objectionable than commitment for dangerousness. Most of our civil commitment systems incarcerate many individuals whose mental illness is such that, although they do not harm others or themselves, they do present others, including their families, with a nuisance. The preferred justification of commitment in these cases is that it is for the good of those committed. When we compare these cases to cases of the commitment of homicidal maniacs or sexual psychopaths, it is easy to conclude that civil commitment on paternalistic grounds is more morally troublesome than commitment for dangerousness, and it is, given the practices in most states. But if the range of harms for which one might be civilly committed were comparable, whether the threatened harm is to self or others, then the difference in moral concern vanishes.

Second, when we think of civil commitment we usually think of a total institution, a warehouse for the mentally infirm in which there is little opportunity for a constructive and joyful life. We can justify placing a homicidal maniac in such a place because custody and constant surveillance and supervision are needed to prevent the harm that is threatened. But when we place persons in such institutions because they are a nuisance and claim that it is for their own good, we are certainly in great moral trouble. The incidence of their nuisance behavior is lessened, but at great cost. The moral problem here is that the moral injunction to use the least possible restraint has not been observed, granting that some restraint on their behavior is morally preferable to toleration. But since the nature of the restraint used in civil commitment is a contingent matter, this feature of current practice does not present a moral difference between paternalistic commitment and commitment for dangerousness as such.

On Paternalism and Health Care

Glenn C. Graber

Department of Philosophy, University of Tennessee, Knoxville, Tennessee

I. WHAT IS PATERNALISM?

I.A. The Model

The model called to mind by the word 'paternalism' is the way a father acts towards his children. Of course, we should not think in terms of a malicious or self-centered father, but rather of a benevolent and loving father who does only what he sincerely believes is in the best interests of his children. Generalizing from this model, we can say that a person is acting paternalistically whenever she or he does something for or to another person without that person's consent and on the basis of the justification that "this is for your own good."

I.B. The Alternatives to Paternalism

What are the alternatives to paternalistic action? One kind of alternative is an imposition on the subject that does not profess to serve the subject's interests. There are several possibilities within this classification. Perhaps the agent is trying to promote selfish interests (e.g., the self-centered parent), or perhaps he or she is furthering the interests of some third party or parties, or perhaps she or he is even attempting to satisfy some abstract principle (e.g., justice). The other alternative to paternalism (the one I shall concentrate on in this essay) is autonomy or self-determination—allowing the subject to make decisions personally without the external imposition of any constraints that are not consented to.

233

I.C. The Conditions for Paternalism

Paternalistic action embodies two key presuppositions that relate both to the classification of an action as paternalism and to the moral justification of the acts so classified. The agent must believe that both of these presuppositions are satisfied for an act to be properly classifiable as paternalistic at all; and, to be morally justified, the belief must be based upon solid evidence.

I.C.1.

First, the agent assumes either that she or he *knows* better than the subject does what actions best serve the subject's good, or else that the individual subject is unlikely to *do* what is best for himself or herself. *(a)* If one were to assume that the subject knows perfectly well what is personally best, and is assured that the subject will pursue it, then it would be pointless to intervene. The agent would be wasting effort, since intervention would not change the outcome, and thus the act could not properly be called paternalism. (b) On the other hand, if one does not assume to know what is best for the subject, then one's action would again not be paternalistic. Rather, it would be either entirely arbitrary, or one of the varieties of other-regarding imposition just mentioned. (c) If one were to assume that one knows what is best for the subject, but on the basis of demonstrably inadequate evidence, then one's action would qualify as paternalism, but would not be morally justified.

I.C.2.

Second, the agent assumes that he or she is in some special way *qualified or authorized* to act on the subject's behalf independent of consent. (a) Without such a belief, the agent's action would not constitute paternalism—but one wonders how such an intervention into another's affairs might then be explained. (b) Further, if the agent assumes having such authority and is mistaken, then the action—although still paternalistic—is (at best) presumptuous. Suppose, for example, that an acquaintance presented you with a bicycle (along with the bill for it, perhaps) explaining: "It's clear that you need to get some exercise." Even if it is true that you do need exercise, this paternalistic act is unjustified. Acquaintances do not have authority to intervene in our lives in this way.

I.D. Justified Paternalism

This example of the bicycle demonstrates that paternalism is not always morally justified. However, I do not believe it can be denied that paternalism is justified in *some* circumstances. Parents would be fully justified, for example, in stopping their small child from drinking poison. *(1)* The parents know about the effects of drinking poison in a way the child evidently does not. *(2)* The role of parent gives one authority to interfere in the life of one's children in such matters.

The next example is more controversial. I would maintain that university faculty are justified, on paternalistic grounds, in setting up some minimal structure of curriculum requirements. *(1)* The faculty knows the benefits of mastering certain academic subjects and intellectual skills in a way the students do not (and cannot, in advance of the required study). *(2)* The role of teacher confers upon them the necessary authority to interfere in this aspect of the lives of their students.

University faculty would not be justified, however, in requiring their students to wear coats in cold weather or shoes when there are hazards to bare feet all about. Even though they may *know* that these actions would be for the students' own good, their authority as teachers does not qualify them to intervene in these matters. (Kindergarten teachers do, of course, have authority to enforce rules of this kind.) This case and the bicycle example both serve to show that superior knowledge is not enough, by itself, to justify paternalistic action. The necessity of special authority to intervene is a distinct, additional requirement.[1]

---

[1] Many authors fail to distinguish these two requirements. See, for example, Kenneth J. Arrow, "Government Decision Making and the Preciousness of Life," in *Ethics of Health Care,* Laurence R. Tancredi, ed., Washington, D.C., National Academy of Sciences, 1974, esp. pp. 34–36; and Bernard Gert and Charles M. Culver, "Paternalistic Behavior," *Phil. Public Affairs* **6** (1), 50 (1976). Gert and Culver offer a detailed analysis of the concept of paternalism that contains some additional requirements. However, I do not think the elements they add are necessary to qualify an action as paternalism. For example, the parent who stops a small child from drinking poison does not violate any moral rule at all (Gert and Culver's feature 3), nor does that parent assume that the child generally knows what is for her or his own good (Gert and Culver's feature 5). And yet, to deny that this act is paternalism is to sever the concept from its linguistic roots.

II. PATERNALISM IN HEALTH CARE

The question that concerns us here is in what circumstances (if any) health care professionals are justified in paternalistic practices.

Some initial guidance in answering this question can be drawn from the sharp restrictions on paternalism imposed by social and legal structures. We will see that here, too, superior knowledge is not sufficient by itself to justify paternalistic intervention in a person's life.

II. A. Professional Relationship

In the first place, health care professionals are not authorized to intervene in the affairs of a given person unless that person is a party to an established professional relationship with them.[2] For example, if a physician whom I met socially were to advance on me with needle and syringe and proceed to inject an antibiotic without my consent, he or she could be charged with felonious assault or battery—even if it were possible to demonstrate that the injection was of benefit to me.

There is a parallel prerequisite for paternalistic practices on the part of university faculty. I cannot, in general, impose curriculum requirements on students of other universities than my own, or course requirements on those who have not enrolled in my classes. Only if a student approaches me in professional circumstances (or, in the case of compulsory education at the primary and secondary levels, is presented by a parent or guardian at the mandate of the state) am I possibly justified in undertaking paternalistic actions.

Most patients choose their own health care professionals and establish a relationship with them voluntarily and on their own initiative. Thereby they give a preliminary consent to having the physician intervene in their affairs. Of those who are unable to take such a step voluntarily—for example, small children and the mentally incompetent, most are brought into such a relationship by a legally or socially

---

[2]The phrase "established professional relationship" refers to the relationship between a health care provider (acting in his professional capacity) and the patient or client. The best-known example is the doctor–patient relationship; but I use a broader designation here because I am concerned with the whole range of health professionals.

I might also point out that a relationship need not be one of long standing to qualify as "established." What is important is that some concrete act of mutual agreement to initiate such a relationship has taken place. It is sufficient, for example, for me to present myself at the clinic and ask to see the physician, who must then agree to see me.

recognized proxy (e.g., parent or guardian) in an exercise of paternalism on the proxy's part. (Thus, in these cases, any act of paternalism by health care professionals is secondary to that exercised by the parent or guardian.) In some institutional facilities (for example, school and factory infirmaries or welfare clinics), patients are not permitted to choose which health professional to consult; but, even in these circumstances, subjects can generally decide on their own not to consult health professionals at all, or else can elect to see private physicians at their own expense. There are some situations in which health care intervention is imposed without any invitation by either the subject or the subject's proxy—for example, legally required immunizations, quarantines, and military physicals; but these are not pure paternalism. These procedures are imposed as much for the purpose of protecting others in the community as to benefit the subject.[3] The only situation in which paternalistic intervention is clearly permitted in the absence of any explicit act initiating a professional relationship occurs in a life-threatening emergency. (We shall discuss the justification for this exception in section *IV.B.* below.)

II. B. Extent of Authorization

The existence of an established professional relationship is thus a necessary condition for authorizing medical paternalism, except in life-threatening crises. In the teacher–student relationship, it is also a sufficient condition for authority to impose minimal requirements (subject, perhaps, to social constraints parallel to those pointed out in the next paragraph). However, this is not a sufficient condition in the professional relationship in health care. There are further restrictions on paternalism within an established professional relationship. If my own physician were to spirit me away to nonemergency surgery while I was sleeping, she or he could be charged with battery—in spite of my having voluntarily enlisted as the patient of that doctor, and even if the benefits of the surgery could be subsequently demonstrated. The general and preliminary consent that establishes a professional relationship does not give the health care provider blanket authority to intervene in my life whenever the provider thinks it wise to do so. Health care professionals are required to obtain explicit (usually written) consent from the patient

---

[3]There are parallels to all of these points in the teacher–student relationship. Mandatory school attendance laws, for example, are justified as much by the interest of the society in having an educated citizenry as in paternalistic concern for the welfare of the child.

(or from the patient's representative, if the patient is personally unable to give valid consent) for any but routine and minor therapeutic procedures.

II.C. Social Structures

Furthermore, the reason that routine and minor procedures are excluded from the requirement for explicit consent is not, I believe, because consent is thought to be unnecessary here. Instead, the presence of consent is guaranteed in a different way—by the social structures of health care. If I did not want the physician to do anything at all to me, I would not go to the trouble of presenting myself at the office or clinic. If I do not want to take the medications prescribed, I will not have the prescriptions filled. If I object to the regimen advised, I will not follow it. Should the physician advance on me with needle and syringe, it is not as easy to refuse treatment as in the other cases just mentioned; but it is still possible—I always have the option of shouting, "Get away from me. I do not want an injection." And that physician could then be charged with battery if the injection were administered in spite of my shouts.

II. D. Further Restrictions on Consent

Still further, even the patient's explicit consent will not suffice to authorize paternalistic intervention unless it is *voluntary* and *informed* consent. Case law offers guidelines regarding how much and what kind of information about risks and alternatives must be communicated to the patient.[4]

All these elements, taken together, leave precious little room for paternalistic action by health care professionals—if the restrictions are taken seriously.

We must not accept these social and legal structures as gospel, of course. They themselves must be shown to be justified. Some professionals view them as excessive and arbitrary barriers to their legitimate

---

[4] One of the strongest of these guidelines is the following: "No consent is valid unless it is made by a person with legal and mental capacity to make it and is based on a disclosure of all material facts. Any fact which might influence the giving or withholding of consent is material." Board of Regents of the University of the State of New York, quoted in Jay Katz, ed., *Experimentation with Human Beings*, New York, Russell Sage Foundation, 1972, p. 60.

exercise of paternalism. Sometimes, the restrictions are ignored. More often, they are honored begrudgingly and minimally—primarily out of fear of legal reprisals, rather than from respect for their inherent rightness.

III. THE CASE AGAINST PATERNALISM

III.A. A Presumption Against Paternalism

I am convinced that these restrictions on paternalism are morally sound and thus that they deserve to be taken seriously.[5] Indeed, I would maintain that they ought to be extended still further, into a general presumption against paternalism in any health care decision.

III.B. The Role of Value Judgments in Health Care Decisions

In order to see the issue here, we must notice that the medical practitioner's management of a patient's illness rests on value judgments as well as technical decisions of medical science. In prescribing medication for pain in the post-surgical patient, for example, not only must the physician determine which dosage will be effective in relieving the discomfort without risking toxic side effects (essentially a technical problem—although it rests on an assumption that death is a disvalue), but she or he must also make a choice between fundamental values. Any really effective pain-killing drug has the effect of dulling the mind somewhat, and many such drugs are liable to lead to addiction. Hence the decision about how long to continue such medication rests on a choice between the value of relief from discomfort and the competing values of maximum mental acuity and freedom from the danger of addiction. All of these are acknowledged values, but in the situation at hand it is not possible to preserve all of these values at once. Hence a judgment must be made as to which is the most important set of values.

---

[5]I hasten to add that I endorse only the legal *principles* involved. I do not agree with certain of the ways these principles have been applied in recent court cases for medical malpractice.

III.C. Paternalism in Health Care Decisions

Physicians often make this decision themselves, on the basis of their judgments about what may be in the best interests of their patients. (In other words, they practice paternalism.) Their judgments may be humane and carefully considered; but the presumption against paternalism entails that this is not the proper basis for such a decision. There are several related reasons why it is not morally justified.

III.C.1. Imposing Personal Values

First, the judgment is likely to reflect the physician's own ranking of values; there is no guarantee that the patient would rank such choices in the same way. For example, the physician might be motivated to relieve the pain on account of personal kindness, whereas the patient may have a stoic attitude toward pain and prefer to endure it and maintain a clear head. If the physician's decision is implemented in this case, the patient's goals are not acknowledged, and he or she is treated as a means to goals that are not personally chosen. (In this case, the foreign goal is the physician's own desire to alleviate pain.) A similar problem would arise if the physician withheld medication in the interest of preserving mental sharpness when the patient would be more than willing to trade a degree of mental acuity for relief from the discomfort.

III.C.2. Assuming Expertise about Values

Further, the health care professional cannot claim to know better than the patient which of these choices is best. The physician's technical expertise assures superior knowledge about the likely *consequences* of each option—i.e., which values are likely to be realized; but that expertise does not include a derivative right to rule definitively on the matter of which value is to be preferred.

Most philosophers who agree with me on this point would defend it by saying that there is no objective basis for preferring one of these values to the other, a position I do not share. I am convinced that some value judgments are rationally justified, and others are clearly unjustified. Specifically, I believe that a case can be made for preferring the stoic, clear-headed endurance of pain—up to the limits of our strength of character. (I cannot take the time here to present this case in any detail. Suffice it to say that it is tied up with the defense of the value of autonomy offered in section *III.C.3.* below.)

In other words, I am *not* claiming that it is inherently impossible for someone to know better than the subject does what is best for that individual's own good. Rather, my assertion is that the *physician* (or other health care professional) cannot claim any special expertise in this matter.

It might seem that what I am saying here is in contradiction to my earlier claim that university faculty have the expertise to justify imposing curriculum requirements upon students. After all, this decision too rests on a value choice. I think there is a difference in the two cases that resolves any apparent contradiction. The value to be obtained from mastering certain academic subjects is what we might call an *abstruse* value—i.e., one that can be appreciated only by those who have already mastered the subjects and thereby have experienced the result. One who has not had the experience of reading quality literature with critical appreciation cannot meaningfully judge the value of that experience. However, the experiences involved in health care choices (e.g., freedom from pain, clearheadedness) are not different in kind from others with which all of us are familiar. Hence there is no place for professional expertise here.

And, if it is argued that some values that we face in medical situations are different in kind from those of daily life, there is still the point that health care professionals have no unique access to the *experience* of these values to give them a basis for authoritative judgment. For example, the male surgeon may know a great deal *about* the reaction of women to radical mastectomy, but he cannot know firsthand what it is like for a woman to face the prospect of losing a breast. Hence he cannot claim expertise on the value choice involved in the decision whether to undertake the added risk of employing chemotherapy or radiological therapy as an alternative to surgical removal of the breast.

III.C.3. The Value of Autonomy

However, these arguments do not go far enough. They do not rule out all forms of paternalism. In fact, they rule out only actions that would be unjustified on paternalism's own ground—i.e., by reference to the superior-knowledge presupposition (§ *I.C.1.* above). For all I have said so far, it would still be possible to justify allowing the physician to make the decision, as long as trouble was taken to learn the patient's own value preferences and the decision were then made on that basis. (This practice is recommended by Eric Cassell in his recent book.[6] He justifies the element of paternalism inherent in this approach with the claim that the psychological trauma of illness prevents patients from reasoning soundly on their own.)

However, I want to argue that, even if the value-rankings of the physician are perfectly coincident with those of the patient in a given case, there is still a moral reason not to have the final decision made on that physician's own initiative. There is an intrinsic value in being in control of one's own life, a value that reveals itself in concrete situations. Suppose, for example, that the physician considering pain-killing medication were to recognize the patient's stoic leanings and make a decision to withhold medication in acknowledgment of the patient's values. This response is still either paternalistic or pointless, since the decision is made by the physician rather than the patient. And I would further contend that this action is morally unjustified. There is a morally significant difference between *(a)* choosing *for oneself* to face the pain and *(b)* having this decision made *on one's behalf* by another; and the moral quality of the first of these alternatives (which amounts to a sort of courage) is denied to the patient if the physician makes the decision.

My contention is that autonomy and personal responsibility are very important values, and that we ought to promote them to the fullest extent possible. Kant emphasizes this point in stirring language in the following passage:

> Laziness and cowardice are the reasons why so great a portion of mankind . . . remains under lifelong tutelage, and why it is so easy for others to set themselves up as their guardians. It is so easy not to be of age. If I have a book which understands for me, a pastor who has a conscience for me, a physician who decides my diet, and so forth, I need not trouble myself. I need not think, if I can only pay—others will readily undertake the irksome work for me.[7]

I think we are in danger of falling into the trap Kant describes here if we allow wide scope to paternalism in the practice of health care. The only sure guards against it are the very strong restrictions on paternalistic action that are incorporated into the social and legal structures I described in section *II* above. In short, patients must be allowed to be

⁶Eric J. Cassell, *The Healer's Art: A New Approach to the Doctor–Patient Relationship,* Philadelphia, Lippincott, 1976, especially Chaps. 3 and 5.

⁷Immanuel Kant, "What is Enlightenment?" in *Foundations of the Metaphysics of Morals,* trans. by Lewis White Beck, Indianapolis, Bobbs-Merrill, 1959, p. 85.

in control of their own lives as fully as possible. Each decision should be made in full consultation with the patient.

III.C.4. The Duty of Autonomy

Elsewhere, Kant goes even further and maintains that we have a *moral obligation* to respect (and promote) autonomy both in ourselves and in others. This way of putting the matter connects it firmly with the authority presupposition of paternalism (§ *I.C.2.* above). The primary responsibility for protecting life and health lies with the subject per se —not with the subject's physician, spouse, the Surgeon General, or any other person. Hence *no one* has a special authority to make health care decisions on another's behalf. Each of us therefore has a moral obligation to accept responsibility for maintaining our own health (e.g., by watching our diet, getting proper exercise, seeking professional assistance when appropriate, following the doctor's recommendations, etc.); and other people have moral obligations to respect our attempts to exercise this responsibility.

IV. THE LIMITS OF AUTONOMY

This defense of patient autonomy also allows us to draw some sketchy conclusions about the limits of autonomy and the sphere in which paternalism is morally justified.

IV. A. Incompetent Patients

A patient who is not *capable* of autonomy (e.g., infants, small children, mental incompetents) will have to be treated paternalistically. However, there is no special reason to designate health professionals as the authority in this situation, since the decisions to be made are fundamentally value choices. The paternalistic authority should be designated *and* guided by society as a whole after extensive and significant public discussion.

A reasonable choice for an authority is a parent or some other relative of the patient, since this will generally be the person most seriously concerned with the welfare of the patient and/or most likely to know the personal values of the patient. However, no one should be given absolute authority over any patient. If the parents of a small child, for example, are seriously failing to act in the best interests of the child, then there should be a mechanism for taking the decision-making authority away from them and transferring it to someone else.

IV. B. Emergency Situations

In life-threatening crises in which the patient cannot communicate personal wishes and in which the family is unavailable, health care professionals are morally justified in proceeding to act paternalistically. Even though they are not *specially* qualified to make value judgments, they are as qualified as anybody else around.

IV. C. Irrational Choices

If an otherwise competent patient makes a blatantly irrational value choice, I think we would be justified in overriding such a wish and acting paternalistically in that patient's behalf. For example, I would have no hesitation in ignoring the wishes cf a patient who announced in the middle of a surgical procedure that he or she does not wish to have the incision closed. ("I would rather die than to have a needle passed through my flesh.") However, this sort of intervention should be limited to the extreme case. Autonomy should include the right to act in ways other people regard as irrational—else it is not genuine freedom. Hence I think there should be careful monitoring of exercises of paternalism on this ground—perhaps through such a mechanism as a requirement for court orders to authorize intervention.

In all these exceptions, the principle that should guide every decision is the attempt to develop (or restore) autonomy in the patient so that personal responsibility can be assumed in the future. If the patient was formerly competent and expressed certain value preferences, these must also be taken into account.

Throughout this essay, I have deliberately stated my claims in somewhat provocative terms in order to prompt discussion. A more cautious statement of my conclusion is the following: Autonomy and self-determination are exceedingly important values, and even an individual's personal duty. Hence paternalism ought to be kept to an absolute minimum in all areas of life, including health care.

Comments on "Paternalism and Health Care"

Joseph Ellin

Department of Philosophy, Western Michigan University, Kalamazoo, Michigan

Professor Graber urges us to keep paternalism to an absolute minimum in all areas of life. It would be easier to assess this injunction if he had explained more clearly what paternalism is.

I. DEFINING PATERNALISM

Graber begins by saying that the model for paternalism is the way parents act toward their children. Fair enough. But he immediately, and inconsistently, defines paternalism ("generalizing from this model," he says, although his definition actually contradicts the model) in terms of the following three conditions: (1) the person acting paternalistically must do something "to or for" another person; (2) without that person's consent; (3) on the basis of the justification that "this is for your own good."

How many of us do things "to or for" our children without their consent? No doubt often we do; but often we do not, also. In fact, the engineering of consent is an important feature of one form of paternalism, found most clearly in the way parents treat children. I will tell you how I persuaded my son to take piano lessons. I told him that when I was a boy my parents wanted me to take piano lessons, which I did; but I never practiced unless coerced and consequently I did not learn very much. I told him that I very much enjoy playing the piano, but do it quite badly (both of these facts he already knew from his own observations), and that I deeply regretted that I could not play any better. I told him that not only for my own pleasure, but for the pleasure of those around me, I wished I had studied the piano more diligently when I had the chance, as he has now. I told him that if he passed up this opportunity, he would regret it in later life. And then I told him that if he takes lessons, I would increase his allowance by $1 a week. And so he agreed. I believe I was acting paternalistically; indeed, I believe I had a paternalistic *duty* to persuade my son to study the piano. And I would not be at all surprised if most parents believe they have such duties. But that there might be a paternalistic duty to persuade someone to do something is ruled out by Professor Graber's definition of paternalism, since according to it, a necessary feature of an action being paternalistic is that the subject does not consent. In the medical context, the engineering of consent is not only *a*, but perhaps *the*, chief feature of paternalism in practice. As Professor Graber points out, medical procedures performed without consent generally constitute a violation of the law. Thus the doctor's problem is to bring you to agree to what the doctor wants to do. Graber's definition would make this necessarily nonpaternalistic, which is much too simple a view of paternalism.

Admittedly, Graber is in good company in requiring subject nonconsent as a necessary feature of paternalism. Gert and Culver, in a recent paper,[1] say, "for if *A* has *S*'s consent, *A*'s behavior is no longer paternalistic." This distortion is a consequence of using examples such as interrupting the suicide attempt, committing psychotics to mental hospitals, concealing information from sick people, or refusing to turn off life-support systems for patients who request it (all of these examples are found in Gert and Culver). What these examples obscure is the important point that the winning of consent can be a paternalistic act.

Now consider another example of what would be considered paternalism. Imagine a large and successful drug company earning a hefty 20% or more on stockholder's equity. The company treats its employees munificently. Not only are wages, hours, and working conditions above prevailing standards, but employees' benefit programs abound: free bus transportation to and from work; subsidized meals at the excellent employee cafeteria; free medical care for the whole family provided by physicians on the staff, and, of course, free medication; discount vacations available at the company-owned hunting lodge in the north woods, and so on. No one is deceived about the company's

[1]Bernard Gert and Charles M. Culver, "The Justification of Paternalism," read at a conference on *Philosophy, Law, and Medicine* at Western Michigan University, October 15–17, 1976.

motive in providing all this: it is using its wealth as tax deductible expense to prevent the formation of a union, which it fears will cause time-loss owing to strikes, will interfere with management prerogatives, and will bring bad publicity, law suits, and government intervention. In this strategy it is totally successful. Though the employees have never asked for any of these benefits, they are happy to receive them, and consider themselves fortunate to work for the company. What makes this company paternalistic, as the employees themselves consider it to be, is that benefits such as these, which employees normally must fight for, are here bestowed as a free gift: the employees have not earned, worked for, or even requested them. But this is not to say that they do not consent to receive them. And so a second point about paternalism that philosophers' examples typically obscure is that the conferring of unearned or undeserved benefits, when these benefits are typically things which must be earned or demanded, is also paternalistic, quite apart from any question of consent.

"Consent" in fact is an inadequate term to catch even the meaning in the standard examples. The expression "without a person's consent" can mean "against the person's will", or it can mean simply that the person has not consented in fact. Unconscious people often have not consented to various procedures which have been done to them, but presumably they would, or might, consent were they able to do so. The would-be suicide, on the other hand, is rescued against his or her will. The problem is that the expression "without a person's consent" in the sense of "has not consented in fact," covers gifts and other unsolicited acts and benefits that typically would not be considered paternalistic at all, for example, Graber's exercise bicycle, which is given as a gift, not as an act of paternalism. (Graber considers this gift unjustified, so that presumably his view is that it is wrong to give anyone a gift unless the gift is harmful or useless.) What makes it paternalistic to administer drugs to the unconscious person, but not to give bicycles to fat people is that the drugs are being forced on the person, but exercise is not being forced on fat people. It is this sense of being forced to accept something that we fail to capture when we use the term, "without consent." That is why the mere giving of a gift, even if done without consent, cannot be paternalistic. Of course, if you then twisted the arms (figuratively) of fat people to force them to use bicycles, you would be paternalistic, not because of any lack of consent, but because the consent is being coerced and not freely given.

There are other elements in Graber's definition that invite criticism. When Graber says the agent must do something "to or for" the subject, he is thinking of obvious examples, such as the doctor who

makes a decision that the patient ought to make. The decision is made
"for", that is "on behalf of," the patient. ("On behalf of" is a phrase
Graber sometimes uses.) To do something for someone means to act as
the person's representative: to act as that individual's agent, in the legal
sense of "agent." But "on behalf of" paternalism is not the only kind
of paternalism, as reflection on engineered consent shows. The doctor
who talks the frightened patient into having a needed operation is, to
be sure, doing something "for" the patient, but in the sense of "doing
for" which means doing something *for the good* of someone. The doctor
is evidently not acting as the patient's representative: that is, the doctor
is not doing something that the patient could just as well do. Since all
paternalism involved "doing for" in the sense of doing *good* for, and
since it is easy to suppose, once "for" is introduced, that the other sense,
"doing on behalf of," is included, it is important to keep these two
senses clear and separate, or one is likely to assume without noticing
it that all paternalism is of the "on behalf of" kind.

Graber's third condition is that the action must be "for the subject's good." This simple phrase masks very complex ideas. What is the
relation between the action being for the subject's good, and the action
itself? It is not necessary for the agent to be motivated by a desire to
do good for the subject: our drug company example shows that paternalistic action can be motivated by self-interest. Must the agent *believe*
that the action is for the subject's good? Not necessarily: it is possible
that the managers of the drug company secretly believe that the employees would be better off with a union and without all the free
benefits. What the agent thinks does not seem a necessary part of the
definition. This would certainly be the case if the secret belief of the
company managers is wrong, and in fact the employees really are better
off with the benefits: in that case the company would have succeeded
in benefitting its employees without either wanting to, or believing it
had. But suppose the managers are correct, and their actions do not in
fact benefit the employees, though the employees think they are better
off. (The arguments of the hapless union representatives who try to
organize the plant fall on deaf ears.) Would we still say that their
actions are paternalistic? Perhaps at this point we would not know what
to say. In our drug company case, the motivation of the managers is
self-interested. They are not acting out of any love for the employees,
but neither are they acting out of malice towards them: their motivation
is neutral in that they do not care whether the employees are benefited,
as long as the employees believe they are benefited. Neutral motivation
is no obstacle to the act being seen as paternalistic. When motivation
is malicious, however, the act cannot be paternalistic: imagine a termi-

nally ill person who actually is better off dead; but suppose that person's enemy, not realizing this, murders the patient from hate. Even though the murderer succeeds in benefitting the patient, no one would call the action paternalistic. On the other hand, if the motivation is paternalistic, the act is paternalistic, even if it leaves the subject worse off, as all sorts of examples, medical and otherwise, will show.

Actually, there are 27 possible combinations of the three factors we are considering, since each of the three factors (motivation, belief, and result) has three values, positive (pro-subject), neutral, and negative (anti-subject). Presumably, many of the combinations do not occur in reality, but we would need to examine each separately to determine which are real possibilities and, among these, which possibilities are paternalistic. A table (p. 250) will clarify the point.In the Table, case 1 represents the agent who is motivated by a desire to help, thinks he or she is helping, and is actually helping: the result is clearly paternalistic. Case 10 represents the drug company whose motivation is neutral (self-interest), but which thinks it is helping, and is in fact helping; case 16 represents a similar company that secretly believes it is hurting, although it is actually helping, whereas case 18 represents the same company when it is actually hurting. Case 25 represents the person who wishes to harm an enemy, believes she or he is harming, but is actually helping. Some theoretical possibilities will be eliminated by the implausibility of a divorce between motivation and belief; it is difficult to think of a case in which a person wants to help someone yet believes the action undertaken is harmful. There are twelve such cases (4–9 and 19–24) and these may be eliminated. Of the remaining fifteen, the six in which motivation and belief have the same sign are clear: cases 1–3 may be paternalistic, but 25–27 are not. The 9 cases in which the agent's motivation is neutral (let us attribute the actions in these instances to self-interest) may be disposed of as follows. If the agent believes help is being provided the subject (cases 10–12), the action may be paternalistic regardless of its actual outcome. If there is no belief that the subject is being helped, the action cannot be paternalistic unless the action succeeds in helping despite the intention (cases 13–16). Thus the result of studying the relation between an action and the action being for the subject's own good is that there are eight cases (1–3, 10–12, 13, and 16) in which the action may be paternalistic. Whether it actually is paternalistic depends on whether other conditions of the definition are met.

Graber adds a fourth requirement to his definition of paternalism. According to it, an action cannot be paternalistic unless the agent is assumed to be "qualified or authorized to act on the subject's behalf." Graber calls this a "key presupposition," and makes it figure promi-

Relation between Motivation and Belief of Agent,
and the Result of Action[a]

Case	Motivation	Belief	Result
1	+	+	+
2	+	+	N
3	+	+	−
4	+	N	+
5	+	N	N
6	+	N	−
7	+	−	+
8	+	−	N
9	+	−	−
10	N	+	+
11	N	+	N
12	N	+	−
13	N	N	+
14	N	N	N
15	N	N	−
16	N	−	+
17	N	−	N
18	N	−	−
19	−	+	+
20	−	+	N
21	−	+	−
22	−	N	+
23	−	N	N
24	−	N	−
25	−	−	+
26	−	−	N
27	−	−	−

[a] + represents a pro-subject orientation; N stands for a neutral orientation; − represents an anti-subject orientation.

nently in Section II of his discussion, but he nowhere explains what "authorized to act" means, or why it is necessary. Obviously anyone who does anything assumes that he or she is "authorized," that is, that he or she has a right to do that thing. But by "authorized" Graber appears to mean something like having a special role, position, or relationship to the subject. It seems then, that on his view, a stranger who rescues a person from suicide, or an adult who prevents someone else's child from taking poison, is not acting paternalistically, since that person has no special role or relationship that authorizes action. But although the authorization condition is put forward as part of the definition of paternalism, Graber actually uses it as part of the justification: the doctor who advances on Graber with needle and syringe is not justified in giving him the injection because he is not in any "authorized" relationship to him; that is, is not Graber's chosen doctor. In addition to this confusion about whether the authorization condition is a condition of the definition or the justification of paternalism, Graber also is confused about whether authorization is a condition of the justification of paternalism, or of the act itself. Graber seems to want to say both that the unauthorized act is not an act of justified paternalism, and that it is not justified, *simpliciter*. Even if it were true that an act is not an act of justified paternalism because the agent is not authorized to do it, it does not follow that the act is not justified: it is simply not justified as paternalism, but might be justified as an act of humaneness, benevolence, mercy, and even justice. If it is Graber's view that the action of a parent in preventing its child from taking poison is an act of justified paternalism because of the parent's special relationship with the child (which is what he appears to say), then presumably the same act of a stranger, if justified at all, must be an act of justified humaneness (or whatever); and the same might be said for the doctor who administers to sick Graber a needed injection, although Graber plainly thinks he has shown that the doctor's act is not justified at all. If, on the other hand, Graber wants to hold that being in the special relationship is necessary for the act to be justified *at all*, then he will have to conclude that the stranger who prevents the child from drinking poison is not justified—which is ridiculous.

It is true that Graber's actual definitions refer not to authorization, but to belief, but this does not rescue his authorization condition. For it follows from Graber's definitions that if a person does not believe he or she is in a special relationship to the subject, an act cannot be paternalistic, whether the person is in the relationship or not. Thus, if the parent does not believe that he or she has any special role-related authority to interfere with that child's life, the interfering act cannot be paternalistic. Similarly, the advancing doctor cannot be about to do something paternalistic, since the doctor knows perfectly well that, not being Graber's chosen doctor, the necessary special relationship does not exist. Many will find this too restrictive. And, again, if something is not a paternalistic act, it follows that it is not a justified paternalistic act. But it does not follow that it is not a justified act.

II. JUSTIFYING PATERNALISM

I have not offered any theory of my own about the definition of paternalism—such an attempt would be out of place in these comments—but I hope that what I have said in criticism of Graber's theory will indicate that paternalism is not the simple, almost transparent concept Graber apparently believes it is. I turn now to the question of the justification of paternalistic actions, however they may be defined.

Graber has two basic reasons for his conclusion that paternalism should be limited as much as possible. He says that agents do not have the same values as subjects; and he says that there is a moral obligation and an intrinsic value to be "in control of one's life". The first reason, however, shows at most that agents should not practice paternalism unless they share, or at least understand, the values of their subjects. This obstacle is probably not as great in the medical context as it is made out to be: patients in general do not want to die, do want to be relieved of suffering, do not want to live their lives as cripples or in great pain. It is true that doctors do occasionally do outrageous things, such as talking a woman into a hysterectomy as a cancer-preventing measure when the unstated motivation is the doctor's belief that she has had enough children already, so that no harm will be done; but morally outrageous cases do not make good grist for the philosophical mill. In other contexts, however, the proviso that paternalistic agents adopt or understand the values of the subjects is inappropriate: the parent who wants its child to take music lessons is trying to change the values of the child, not to reflect them. Fostering the development of values is one of the aims of paternalism, whether the context is one of psychological, moral, or religious counselling, or of the development of taste in the arts, or of raising the level and tone of political debate. A paternalistic government, for example, might provide subsidies for culture in order to foster appreciation of esthetic values among the population. A candidate for President might make it a campaign goal to raise the level of political debate in the country. These actions would be paternalistic, even according to Graber's definition, and what is wrong with them?

Actions of this kind ought to be encouraged, not kept to an "absolute minimum."

Graber's main reason for opposing paternalism is that autonomy is an intrinsic value and we have a moral obligation to control our own lives. The first thing to be said here is that the opposition between paternalism and autonomy is misstated. The kind of paternalism a parent exercises toward its child—the paradigm of paternalism, according to Graber—does not hinder autonomy, but fosters it: encouraging a child to stand on its own feet is what parenthood is all about. The point here is that we do not necessarily help children become independent by assuming that they already are independent: the parent must at each stage of the child's development be sensitive to just how much independence the child is capable of handling, and then encourage the child to reach for that much, and perhaps a little bit more. Since autonomy is a matter of degree, the same principle applies to adults. We might well believe that an adult who has a keen critical mind, an appreciation of fine art, and a taste for sophisticated pleasures, is a more independent, more self-directed, more free person, than one who does not; thus the program of raising the level of the public's taste and intellect might very well be paternalism carried on in the interests of greater autonomy for everyone.

But let us consider the more conventional case, where doctors or other professionals make decisions for people in trouble. Graber is against this because "autonomy and personal responsibility are very important values, which we ought to promote to the fullest extent possible." This argument is a *non sequitur:* a person who wishes to promote something as much as possible only because it is an important value is a fanatic, not a philosopher. We would want to promote autonomy as much as possible only if it were the only value, or the overriding value; there are other important values besides autonomy. But that autonomy overrides all other values is exactly what Graber must prove (not assume) to make out his claim that paternalism ought to be kept to a minimum. Another value, for example, is the relief of suffering and alleviation of anxiety. So is good health generally. If a doctor practices a little paternalistic deception (overemphasizing the dangers, perhaps) to persuade someone to quit smoking or go on a diet, I do not see what is wrong with this. Right now the automobile industry is practicing a bit of paternalism (out of self-interest to be sure) in order to persuade people to fasten their seat belts: there are commercials calculated to induce guilt over the suffering you will have caused your loved ones if you should be killed as a result of driving with the seat belt unused. Now I happen to feel just a bit offended at these advertisements, but this is because their emotional appeal is so blatant, and their ultimate purpose so transparent, not because they are paternalistic. If the government were to sponsor such advertisements, and if by so doing it were substantially to reduce traffic fatalities, the paternalism involved, it seems to me, would be a small price to pay. In fact, I fail to see why we should not require the government to take some such action, if traffic injuries amount (as some people say) to a national health emergency: is there not a paternalistic obligation on the part of the government to devise measures by which people may be persuaded to do things for their own good? The subject of paternalistic obligations requires more philosophical analysis than it has received; it surely cannot be disposed of by waving the flag of autonomy.

The case against paternalism, I conclude, is at best not proven.

V

PROFESSIONAL
RESPONSIBILITY

22

Teaching Compassion: Professional Education for Humane Care

Arthur J. Dyck

Mary B. Saltonstall Professor of Population Ethics,
Harvard University, Cambridge, Massachusetts

It is one of the ideals in medicine to be known as a compassionate physician. Yet medical students often complain that they enter the study of medicine with a strong desire to be kind to others, and during their medical education find themselves losing this ideal. Apparently the rewards of medical education are gained largely by becoming a technically competent person, and not in becoming a kind person.

This concern of medical students is shared by many of their teachers who ardently desire that love of the art of medicine be coupled with compassion. But many well-intentioned analyses of how compassion is to be nurtured in medical school both reveal and exacerbate the situation in which medicine as a technical skill is well learned and compassion loses ground. For example, in a report first written for the Dean of Medicine at McGill University,[1] Dr. Donald Bates uses the word humanism to describe an attitude of love for human beings. This "humanism," which includes what we are calling "compassion," is for Bates a personal value. It requires direct and protracted contact between the individual student and a particular teacher. The best opportunity for fostering compassion occurs, according to Dr. Bates, during clinical instruction:

---

[1] Donald G. Bates, "Humanism in Undergraduate Medical Education," *Canadian Medical Association Journal* 105 (1971), pp. 258–61.

... difficult though it is to arrange, the student's early clinical experience needs to be combined with an exposure to those who have been successful, not only as doctors, but as human beings.[2]

Notice here how Dr. Bates, I think quite unintentionally, allows for the separation between being a successful doctor and being a successful human being. This separation creates the second-class role for compassion which so bedevils efforts to instill it; compassion is important, but it cannot match the importance of everything else that makes for success in medicine; if success is possible without compassion, it is dispensable.

One may agree with Dr. Bates, as I do, that compassion is an attitude or kindly disposition that can indeed be learned in the way and in the context that Dr. Bates has suggested. However, despite the existence of these opportunities to convey compassion, we are still confronted with the stark reality that many medical students fear their education has eroded their compassionate impulses. And patients looking for compassion all too often find physicians wanting in this regard.

If, then, one takes the position that there is some need to teach compassion, or at least to encourage rather than discourage it, why not do this by inserting the humanities, particularly ethics, into the medical curriculum? At first blush, this seems to be an eminently reasonable suggestion, but there is a startling and strong objection to such a move from within the community of humanities itself, spearheaded by people in ethics.

Robert Veatch and Sharmon Sollitto, in a report concerned with teaching medical ethics in medical schools, raise the question whether medical ethics can indeed ever be taught.[3] Doubt about the possibility of teaching ethics will be expressed by those like Bates who say the physician should learn medical ethics by osmosis while sitting at the feet of other physicians. But the impossibility of teaching medical ethics will be asserted by those who "perceive medical ethics as a source of sensitivity training that is expected to produce warmer and more compassionate physicians."[4] What Veatch and Sollitto contend is that if you see medical ethics in this way, then it cannot be taught:

Anyone beginning a medical ethics teaching program with the promise that the physicians produced by that program will be warmer and more sensitive probably is in for a major defeat.[5]

---

[2]*ibid.*, p. 260.

[3]Robert M. Veatch and Sharmon Sollitto, "Medical Ethics Teaching: Report of a National Medical School Survey," *JAMA* **235** 1030–1033 (1976).

[4]*ibid.*, p. 1032.

[5]*ibid.*, p. 1032.

To a certain extent, I have to agree with the alliance here between the
views of Bates on the one hand, and Veatch and Sollitto on the other.
It is surely true that personal relations and the compassionate practice
of medicine are potentially the best sources of inducing compassion.
And a course in medical ethics in turn will not be as dramatic a source
for the illustration of actual kindness in medicine. However, there is
more to compassion as an attitude than meets the eye, and there is more
to be said about compassion than we learn from seeing it as an attitude.

In a recent report on the teaching of bioethics under Veatch's
editorship, the argument goes beyond the claim that compassion is
difficult to teach: compassion should not even be part of the explicit
subject matter of ethics or any of the other courses in the medical
curriculum.

> The special purpose of bioethics teaching should not be to increase
> humanitarianism. If such an increase results at all it derives more gener-
> ally from the educational experience and the role models for life styles and
> value perspectives that students see in their teachers and friends. Presum-
> ably every teacher hopes that students will leave a course changed in their
> ability to think, reason, or act—and this applies also to courses in the
> humanities, including bioethics, but bioethics is in no way unique in this
> hope.[6]

This Hastings Center Report does have an important point to
make. Certainly bioethics should not be the only subject concerned with
evoking compassion. It should not in this respect be unique. Compassion is the concern of everyone and of every course. But is there no
special contribution that the subject matter of ethics can make to understanding what compassion is, to leading people more highly to value it,
and hence also to providing a special opportunity to incline people
toward its cultivation? But here our conception of the subject matter
and the purpose of ethics becomes critical. If one insists, as does the
philosopher Danner Clouser, first, in agreement with the Hastings
group that ethics has no special reason to be concerned with compassion; second, that students of ethics are developing skills (not motives)
for sorting out confused and complicated ethical issues; and third, that
students who choose to do what is morally right are "influenced far
more by the example of their medical mentors than by anything they
could learn in an ethics class"[7]—the hope that more compassionate

---

[6] *The Teaching of Bioethics,* The Hastings Center, Institute of Society, Ethics and
the Life Sciences, 1976, p. 9.

[7] K. Danner Clouser, "Medical Ethics and Related Disciplines," in Robert M.
Veatch, Willard Gaylin, and Councilman Morgan, eds., *The Teaching of Medical
Ethics,* New York, Hastings Center, 1973, p. 40.

physicians should be expected to emerge from classes in ethics is indeed misplaced.

Behind the disclaimers regarding the role of ethics in teaching compassion is a particular view of compassion as well as of ethics. Compassion, says Clouser, "has nothing to do with the study of ethics (though it might motivate one to study ethics) and it is not even a moral obligation. Insofar as compassion is a subjective state or feeling, it could not be required by ethics."[8]

I disagree with Clouser on both counts. First, because compassion is an attitude, but it is more than that. Secondly, because compassion as an attitude is a moral obligation and it is obligatory in some of its other forms. In developing these two observations in the remainder of this essay, I wish to indicate my thinking about what compassion is, why it is obligatory, why it is possible to teach, and why ethics has a special though not unique opportunity to teach it.

I. THE NATURE AND SCOPE OF COMPASSION

In dictionaries, compassion is portrayed as an attitude with two components: (1) deep sympathy or pity for the suffering or trouble of others; (2) deep sympathy accompanied by an urge to help. Compassion, then, is a feeling of pity and a disposition that inclines us to act in ways that are helpful. Sympathy or pity also inclines us to refrain from injuring others or causing them suffering. What can and should ethics teach in order to encourage rather than discourage compassion?

I.A. Compassion as a Moral Virtue

There is general agreement that improving one's own character is a moral obligation. Among the improvements of character of the utmost significance for the moral life are the ability to perceive when actions are right or wrong, and the willingness to do what is right. Pity as a disposition or capacity to feel how it is for others, to feel sympathy when others are suffering, is a necessary ingredient in enabling us to know when it is that people are suffering or otherwise in need of help.

It should come as no surprise, therefore, that recent discussions in ethics include the ability vividly to imagine how it is for others as a

---

[8] *ibid.*, p. 42.

requisite to making reasonable moral judgments.[9] Moral judgments made without a lively sense of pity (sympathy) that allows one to feel how it is for others may lead us to violate some of our most stringent and important moral obligations.

Improving ourselves with respect to sensitivities that help us discern what is right or wrong is not enough to insure that we will meet our usual moral obligations. It is equally important that we have the will to do what is right. Successfully to refrain from injuring others depends both upon being able to sense what would be injurious and upon being willing to see others free from any injuries. Recently an example of teaching compassion came to my attention.[10] It seems that physicians sometimes ask young adolescents to undress in front of their parents. This is very distressing to these adolescents. Yet many physicians do not think about this. When this source of unhappiness for their patients is called to their attention, physicians are quite eager to protect adolescents against it. Here the pity of one physician, and his willingness to teach what he discerns as a result of it, encourages and evokes similar feelings and behavior in other physicians.

I.B. Compassion and Doing No Harm

In the example just discussed above, we have illustrated how compassion is necessary to perceiving and desiring to do what is right. Ethics can and should teach this. The example above illustrates something else that ethics can and should teach. There are certain actions that are congruent with compassion and others that are incongruent. In this respect, the vocation of physicians and other health professionals is made immensely difficult by the simple but momentous fact that they must often inflict pain, injury, and suffering in order to satisfy their urge to help. They must overcome the usual inhibition pity induces against injuring a person, as they must, for example, in cutting open patients in surgery. But, paradoxically, in order to do this, they must retain or generate sufficient sympathy for the suffering of patients which would result if they failed to overcome this inhibition against injuring them. In other words, health professionals must have a lively sense of the evil

---

[9]The best philosophical discussion of these criteria is found in Roderick Firth, "Ethical Absolutism and the Ideal Observer," *Phil. Phenomenol. Res.* **12**, 317–345 (1952). See also Arthur J. Dyck, *On Human Care: An Introduction to Ethics,* Nashville, Abingdon, 1977, pp. 141–155 for some of the practical applications and implications of Firth's theory.

[10]Dr. Melvin Levine—a personal communication.

they are removing or preventing so that they can live with the lesser injuries that medicines or surgery inflict on their patients. At the same time, health professionals need a healthy dose of pity so that they will not in any way harm their patients for the sake of money, or harm them in those instances in which the present suffering of the patient is less than what would result from medical intervention. Medical intervention that is congruent with compassion lives under the strict obligation first of all to do no harm, and secondly, to inflict injury only for the sake of removing or preventing a greater injury.

Ethics can and should remind health professionals and others of what an important guide to our thinking about right and wrong the element of pity really is. Rape, torture, terrorism, and the like, are not actions done out of a deep sense of pity, but are actions that most of us recognize as pitiless. As recent studies have shown, violence perpetrated in a context that does not elicit sympathy for the victim influences people toward violence.[11] Thus, for example, someone portrayed as completely deserving the beating received in a filmed fight sequence encourages those who have viewed the film to engage in behavior that brings pain to others. But if those who view this same film are told that the victim was carried to the dressing room unconscious and died there, the violence is seen as excessive, and this sympathy for the victim inhibits subsequent aggression on the part of these viewers.

I.C. Compassion and Justice

Bernard Barber has recently suggested that compassion in medicine be understood as a form of egalitarian relationship between physicians and patients.[12] He does not dwell on compassion in any of the senses we have already discussed, but focuses entirely on its link with equality. He urges that medicine train its practitioners to treat patients as equals, as individuals capable of making informed decisions, and as bearers of values whose judgments are to be respected. He does not claim that physicians and patients can or should be equal in every respect, since physicians will bring to the physician–patient relationship an expertise one cannot expect patients to possess.

Barber believes that the teaching of ethics will help to encourage

---

[11]Leonard Berkowitz, "Impulse, Aggression and the Gun," *Readings in Social Psychology Today,* Del Mar, California, CRM Books, 1970, pp. 89–93.

[12]Bernard Barber, "Compassion in Medicine: Toward New Definitions and New Institutions," *New England J. Med.* **295,** 939–943 (1976).

more egalitarianism, but he does not in this particular essay give any recognition to compassion as a moral virtue. The achievement of a more egalitarian physician–patient relationship will certainly depend on cultivating compassion in the form of a willingness to do what is right and an ability to know what is right. The respect for the autonomy and values of patients will need to take the form of recognizing obligations to tell the truth, to keep promises and honor contracts, and to treat people as equals. Without the acceptance of these moral principles and without the willingness to honor them, egalitarianism will be a mere abstract goal, the content of which will be relatively empty, and the practice of which will be virtually nil. Thus, for example, I have no quarrel with Barber in insisting on written informed consent that includes the explicit understanding that patients may always break off their relationships with specific physicians, but what it is that patients think they are signing and the extent to which they feel it is best to comply with physicians depend very heavily on the integrity, conscientiousness, and sympathetic manner of the physicians who seek the consent of patients. The most impeccably written consent forms cannot completely protect patients from those physicians who fail to convey through their words, actions, or gestures, the true extent of any risks of real pain or loss of life that may be entailed in a therapy or experiment for which consent is being sought.

Hence, although Barber does contribute an important insight into one of the facets of compassion, a formal recognition of the liberties and rights of patients will not suffice to protect them against the more subtle, often unintentional ways in which physicians obtain from patients what physicians think is best for them. One of the most important protections against mistakes in moral judgment is a genuine pity for the condition of the patient, and hence, of any genuine risks that may be incurred by medical interventions. Similarly, the willingness to abide by those moral principles that guide and make an appropriate amount of egalitarianism possible will rest, as we have already indicated, on the extent to which physicians are morally conscientious and morally perceptive.

I.D. Compassion and Reason

There may be those who are thinking that all of this is very interesting, but still unconvincing, and perhaps even dangerous. The idea that health professionals or others ought to cultivate the emotion of pity and be guided by it may suggest to some the introduction of an irrational element in moral decisions. After all, are not moral decisions to be decisions that are impartial, that is, in some sense disinterested and dispassionate? This is true, and so it is important to discuss why it is that compassion, though one important element in the proper cognition of what is right and wrong, is not the whole of what makes for reasonable moral judgments.

There is no space here to treat the various theories of a metaethical sort that discuss the manner in which and the extent to which our moral judgments may be rendered more rational or justifiable. Readers need to look elsewhere for such an important discussion.[13] Suffice it for our purposes to state that there are at least three criteria of rationality that are widely recognized in ethics, namely, being factually informed, vividly imagining how what we are contemplating will affect others, and impartiality. One way to see how emotions contribute to reasonable moral decisions and how also they are kept in check by other elements of cognition is to examine the court procedures we have developed for achieving justice for contending parties.

Consider the court procedures connected with trying someone alleged to be guilty of a major violation of the law. In order to assure impartiality in one of its important senses, we require that jurors be screened for any obvious prejudices they may have with regard to the person on trial or with regard to publicity about the alleged crime. The judge must also be impartial in the sense of being free of any conflict of interest. Here we are insisting that a judge not be partial in the way that judges trying their own children would be. We expect, furthermore, that all the relevant facts pertaining to guilt or innocence be brought out, and for this purpose, we use prosecuting and defending attorneys. This practice has an important emotional component, an element of compassion, in it: The facts are to be brought out by persons (lawyers) who are thoroughly and sympathetically committed to the parties for whom they contend. We attempt to assure that lawyers will give a vivid and sympathetic account on behalf of their respective clients so that jurors and judges will vividly imagine how it is for these clients. Courts are not content to have a single person represent both contending parties in a trial. There is a recognition of the need for a thorough and sympathetic portrayal of how it is for each of the contending parties. Courts embody both the notion of freedom from distorting passions and positive use of sympathy in still another of their

---

[13]See, for example, chapters 7–10 in Richard B. Brandt, *Ethical Theory*, Englewood Cliffs, New Jersey, Prentice-Hall, 1959. See also chapters 6 and 7 in Dyck, *op. cit.*

practices. Increasingly courts are guided by the notion that fairness will be best served if cases involving blacks and women are not judged by juries composed entirely of white males. By assuring black and female presence when appropriate, a subtle kind of sympathy and understanding, and of imagining how it is for someone, is surely introduced into the endeavor to achieve a more rational and fair verdict for persons who are being tried.

The courtroom provides but one example of how we seek to determine what is just by the fairest and most rational procedures we are able to discern. These court procedures reveal very well how emotions have a definite, though limited, role in resolving disputes and achieving fairness in a way deemed to be reasonable.

II. CONCLUSION

We have argued that compassion can and should be taught to medical students and practitioners. Although it can and should be taught in many different settings, ethics has a special obligation to teach it for the following reasons:

1. Compassion as a sense of pity and urge to help is a moral virtue. This moral virtue encompasses both an ability to discern what is right and wrong, and the willingness to do what is right.

2. Ethics as a discipline can offer cogent and systematically considered reasons why compassion is obligatory.

3. Ethics can identify the relationship between compassion and other moral obligations.

4. Ethics can explain how compassion contributes to rationality in cognition and can help counter fears that compassion of necessity introduces an irrational element in moral decision-making.

I do not expect that everyone will agree with this account of compassion. But I would hope that readers would be persuaded that there are important reasons for being compassionate. The injunction "first of all to do no harm" is a time-honored constraint against unnecessary or otherwise unwarranted injury. The vividness with which it is recognized and the extent to which health professionals comply with it will derive in no small measure from the cultivation of the compassion that makes this constraint a reality. Success in medicine demands this.

But the pity that constrains health professionals and others from inflicting injury must be imaginative enough to move them to the prevention and removal of pain, suffering, and risk of death by means of actions that will themselves entail pain, suffering, and risk of death. Success in medicine demands this also.

Compassion is at the heart of medicine in two more, very significant ways. Properly to diagnose illness requires a great deal of sympathy and patience, an atmosphere that encourages free and open discussion of the hurts that bring patients to physicians. Comfort and hope for patients are essential to therapy and are encouraged by compassionate care. Ethics may confirm, but medicine must teach how indispensable to success in medicine are these two modes of compassion.

Accountability in Health Care Practice: Ethical Implications for Nurses

M. Josephine Flaherty

Principal Nursing Officer, Department of National Health and Welfare, Ottawa, Ontario, Canada

Whenever health care professionals meet, they express concern about the number and complexity of ethical dilemmas that they face and the effect of these on the quality and quantity of their professional practice. As consumers become more aware of and troubled about ethical issues, it is crucial for health professionals, who face difficult personal and professional value conflicts, to be able to identify and articulate thoughtful ethical positions as individual human beings and as professionals.[1] Ethics in health care is not something to be restricted to pieties in conference rooms or seminars. Rather, it is a necessary, useful, and productive basis for action.[2]

At least since the time of Hippocrates, people have recognized that morality has a central place in the practice of medicine. Moral codes and ethical theory have a long and honored place in both philosophical and medical literature. Yet until this century, the domain of this subject has been limited largely to physicians' professional obligations to each other and to their patients. Expansion of the biological sciences has been accompanied by unprecedented social and cultural changes that are affecting medical structures and decision-making. Human rights and individual autonomy are dominant political themes, authoritarian

---

[1] Mila Ann Aroskar, "Ethics in the Nursing Curriculum," *Nursing Outlook* **25**, 260–264 (1977).

[2] Eric J. Cassell, "Autonomy and Ethics in Action," *New England J. Med.* **297**, 333–334 (1977).

and hierarchical decision-making is being replaced by pluralism and personal freedom of choice, and increasing social concern for personal and environmental health has prompted questions about the value of health in relation to other desired goals in our society.[3]

These developments have helped to create for health science professionals new and different moral dilemmas related to the use of knowledge and technology to affect human procreation, to modify behavior, and to influence the course of death. These are bound inextricably to questions about the concepts of health and illness, the goals of health care, the relationships between patients and those who care for them, and the efficiency and effectiveness of complex institutions such as the health care system.

The field of bioethics attempts to ask the normative questions of ethics as they arise in health care research and practice. As an interdisciplinary field, involving cooperation and discussion across traditional discipline boundaries, bioethics brings numerous perspectives to critical issues bearing on "whole patterns of private, public, governmental and institutional life."[4]

Health care practitioners of the seventies have declared that they are professional and that they want to embrace the privileges and responsibilities of professional status. Probably never before have health care workers been faced with such complex issues and been called upon to make such far-reaching decisions. There are certain professional characteristics that have prepared them for their work.[5]

The first of these is education, both general and specific. General education has provided opportunity for health professionals to learn to think and reason with accuracy, and to develop an appreciation of the world in which they practice their professions; specific education has provided a sound theoretical framework for practice. Educational programs have not been static, but have been modified to meet the needs of students in the light of the professional demands that will be placed upon them.

A second characteristic of a profession is acceptance of a code of ethics. Health care professionals place high value on the worth and dignity of human beings and that value directs them in their practice. As society changes and roles are altered, ethical codes may be subject

---

[3] *The Teaching of Bioethics.* Report of the Commission on the Teaching of Bioethics. Hastings-on-Hudson, New York, Institute of Society, Ethics and the Life Sciences, 1976, pp. 1–3.

[4] *ibid.*, p. 4.

[5] M. Josephine Flaherty, "Professional Obligations with Effective Rewards," *Alberta Association of Registered Nurses Newsletter,* June, 1975.

to modification. The basic quality demanded of an ethical professional is integrity, a definition of which was given to me a decade and a half ago by one of Canada's outstanding medical scientists, who has been demonstrating integrity since before the beginning of this century. To Dr. Peter Moloney of the Connaught Medical Research Laboratories, acting with integrity means doing what one believes to be right, regardless of the cost.[6] Belief in something implies that a person has given careful thought to the information at his or her disposal and has arrived at a logical conclusion. Beliefs may be modified in the light of new evidence. Competent health professionals, whose philosophies are eclectic as they borrow from the ancients as well as the moderns for principles and guidelines, can appreciate the words of Marcus Aurelius who said in *The Meditations:*

> "A man should always have these two rules in readiness; the one, to do only whatever the reason of the ruling and legislating faculty may suggest for the use of men; the other, to change thy opinion, if there is anyone at hand who sets thee right and moves thee from any opinion. But this change of opinion must proceed only from a certain persuasion, as of what is just or of common advantage, and the like, not because it appears pleasant or brings reputation."[7]

The person with integrity, then, has the courage to change his or her mind and to admit publicly that he or she was wrong.

A third characteristic of professionals is dedication to the ideal of master craftsmanship in their work. The realization of such an ideal involves more than simply being able to do things well—it requires knowledge and understanding of the principles on which theory is based as well as the ability to apply these in the practice of the profession. In *The Metaphysics,* Aristotle said that "the master craftsmen in every profession are more estimable and know more and are wiser than the artisans, because they know the reasons of the things which are done;. . . . Thus the master craftsmen are superior in wisdom, not because they can do things, but because they possess a theory and know the causes".[8] True mastery of a profession is not something that is acquired suddenly. Rather, it is an on-going process that demands of each professional that he or she strive constantly to add to personal knowledge, to perfect professional skills, and to enlarge the body of

[6]Personal communication with Dr. Peter Moloney, Toronto, 1962.

[7]Marcus Aurelius, *The Meditations,* IV, 12 (trans., George Long), Garden City, New York, Doubleday, 1960, p. 37.

[8]Aristotle, *The Metaphysics* Book I, 1, 11–12 (trans., Hugh Tredennick), London, Heinemann, 1933, p. 7.

knowledge for the discipline. The hallmark of a master craftsman is an enquiring mind and a commitment to continuous learning.

A fourth characteristic that is part of professionalism is informed membership and involvement in the organized profession. The member who is intellectually self-employed thinks and speaks for himself or herself and acts according to his or her own decisions rather than according to what someone else has told him or her to do. The member with an enquiring mind knows what is going on in the profession and is involved in the development of new patterns. He or she will not tolerate the absolutism that is the result of a tired democracy.

The final characteristic, one which subsumes the other four, is accountability, or the taking of responsibility for one's own professional actions. The true professional does not blame others for what has been done or not done in the profession and in the society in which she or he lives. Rather, the professional participates in decision-making and planning and learns to live with the decisions. He or she accepts the fact that failures will be experienced along with successes, but believes that if he or she acts responsibly, the successes will outnumber by far the failures. The professional strives constantly to practice in a diligent, reasonable, and justifiable manner, to document his or her reasoning, and to display a willingness to subject it to the scrutiny of peers. Thus do health professionals, who no longer feel obliged to carry the burden of omniscience, develop and apply strategies to deal with almost instant obsolescence of knowledge and professional practice and achieve the "Maturity: [that means] among other things—not to hide one's strength out of fear and, consequently, live below one's best."[9]

The nursing profession, which shares these characteristics with the other professions, has been affected markedly by the unprecedented scientific and technological advances of our age and by changes in the delivery of health care. Nursing practice, for which members of the profession are accountable, is thought to be composed of at least four steps or dimensions. These include planning, implementation, evaluation, and research.

Planning involves the recognition of the real or potential problems or needs of patients, and the identification of strategies for coping with these. Planning may be narrowed to one or more individuals or broadened to a department, an institution, or a community. Implementation involves decisive action towards a defined goal. It can become routinized, cold, almost mechanical, but it has the potential to be the most creative, communicative, and satisfying

---

[9]Dag Hammarskjold, *Markings*, London, Faber and Faber, 1966, p. 87.

component of the nursing role. It is the "art of nursing." Evaluation is the process of determining the significance or worth of nursing action by careful appraisal and study. It should involve also the determination of appropriateness. It takes place concurrently with and retrospectively to implementation, as well as prior to new planning. One of the marks of a profession is that it monitors its own practitioners, and thus accountable nurses must be prepared for peer review as well as self-evaluation. Research involves disciplined study of the effects of nursing actions that will lead to the development of nonarbitrary standards for nursing practice, standards that are developed by specialists in nursing rather than by administrative or organizational tradition or convenience.

This professional scope of nursing embraces three complementary aspects: The first is the *independent* function, in which the nurse makes judgments that are based on education and experience and that depend on sound theoretical knowledge. The second is the *dependent* function, in which the nurse acts according to the directions of physicians or other health professionals, and according to policies of health care agencies. The nurse is obliged by ethics and by law to question those directions and policies about which he or she has concern. The third is the collaborative or *interdependent* function, in which the nurse works with patients, families, and other members of the health care team in the effort to meet patients' needs. This requires mutual respect and cooperation among health care workers.

The scope of nursing is that of health care, in the broad sense, as opposed to strictly illness care. Its goal as a field of professional endeavor is to help people attain, retain and regain health. The phenomena with which nurses are concerned are human health seeking and coping behaviors as people strive to attain health.[10] Nursing in this country has committed itself, through practice according to *nursing* models that lend themselves to change and to adaptation to many venues, as opposed to *medical* or other models, to assess the functional or coping levels of individuals and families in the light of their biology, their environment, their life style, the health care system, and the interactions among these, to plan and implement proactive and reactive nursing intervention and to evaluate the effectiveness of this. Were all nurses able and willing to identify and assess human coping behaviors in ways that would facilitate early and accurate detection of present or potential problems in health behavior of individuals and groups, the

---

[10]Rozella Schlotfeldt, "This I believe. . . . Nursing is Health Care," *Nursing Outlook* **20**, 245 (1972).

effect of nursing on the health status of the community could be very significant.

According to provincial laws, to ethical codes, and to standards of professional practice, nurses in Canada are significantly independent professional practitioners of health care who are *accountable for* their behavior, rather than *accountable to* someone in a hierarchy. In the province of Ontario, for example, registered nurses are governed by the *Health Disciplines Act, 1974, Parts I & IV,*[11] by the *Regulations Under the Health Disciplines Act 1974, Nursing,*[12] by the *ICN Code for Nurses —Ethical Concepts Applied to Nursing*[13] that was endorsed by the Council of the College of Nurses of Ontario, and by the *Standards of Nursing Practice: for Registered Nurses and Registered Nursing Assistants*[14] that were adopted by the College of Nurses under Regulation 24. Under these provisions, every registered nurse must exercise "judgement in accepting responsibility and is accountable for his own actions."[15] The acts of registered nurses require substantial knowledge, skills, and judgment, and are performed either independently or in cooperation with other members of the health care team.[16]

In the province of Quebec, the regulation and the autonomy of the nursing profession are recognized in *Bill 250, The Professional Code* and *Bill 273, The Nurses' Act.* In the latter, nursing practice is defined as involving accountability for behavior, independent practice, and dependent practice, which must be carried out within the context of the knowledge acquired, the established customs of the institution, the code of ethics, and the standards of practice of the profession. Other provinces in Canada have similar statutes.

The ethical implication of this for nurses is that they are accountable for professional behavior that involves application of the nursing process and cooperation with other members of the health team, within current legislation affecting the practice of nursing and according to the profession's code of ethics. This gives nurses, with other health professionals, the responsibility to practice with competence and to exercise

---

[11]Government of Ontario, *The Health Disciplines Act, 1974,* Statutes of Ontario, 1974, Ch. 47 as amended by Ch. 63, 1975, Toronto, Queen's Printer, 1975.

[12]College of Nurses of Ontario, *Regulations Made Under the Health Disciplines Act, 1974,* Toronto, College of Nurses, 1975.

[13]International Council of Nurses, *ICN Code for Nurses—Ethical Concepts Applied to Nursing 1973,* Geneva, International Council of Nurses, 1973.

[14]College of Nurses of Ontario, *Standards of Nursing Practice: for Registered Nurses and Registered Nursing Assistants,* Toronto, College of Nurses, 1976.

[15]*ibid.,* p. 5

[16]*ibid.*

judgment in the preservation of the safety, dignity, and autonomy of patients.

Although few people would neglect the preservation of safety and dignity, nurses, like many professionals who attempt to establish practices modeled as closely as possible on the system under which they were trained, are tempted to slip into an Aesculapian model in which goals may be defined *for* patients rather than *with* them. Nurses practicing according to current conceptual models do not sustain hope in patients by promising survival or the cure of disease (that patients often know is not possible); rather, they sustain hope by keeping patients as much in command of themselves, their symptoms, and their situations as possible[17] and thus they preserve the autonomy of patients. Autonomy calls for a degree of independence or freedom of choice for the individual, and requires for that person knowledge, ability to reason, and ability to act in a way that is true to his or her nature. No one is truly autonomous; everyone needs and depends somewhat on others to authenticate his or her acts, to provide knowledge, to assist with thinking, and to facilitate choices. In health care, the relationship between the professional and the patient is a medium by which autonomy can be enhanced. Professionals succeed in the preservation of autonomy when those patients requiring continuing care are able to function with the least possible interference from their disease or their therapeutic regime.[18]

In the seventies, perhaps more than ever before, health care professionals must go beyond direct patient care to consider and act upon factors associated with the nature and shape of the health care system, the responsibilities and practice of members of health care teams and the changing roles of consumers in the maintenance of their own health. Seven years ago, in *Future Shock,* Alvin Toffler warned that "between now and the beginning of the twenty-first century, millions of ordinary, psychologically normal people will face an abrupt collision with the future."[19] Although some people fear that this will result in human organisms being loaded beyond their capacities to adapt, health science professionals have no fear of this, for they and their clients have been adapting constantly and successfully since the beginning of time.

Nursing, as one of these professions, has as its goal to promote the adaptation of clients or patients in situations where changes in health status (including illness) place new demands on those patients. Nursing

---

[17]Cassell, *op. cit.*

[18]*ibid.*

[19]Alvin Toffler, *Future Shock,* New York, Bantam, 1971, p. 9.

prides itself on being a profession that gathers and analyzes data relative to human beings and their health problems, plans with those clients to deal with the problems, mental and physical, taking into consideration those factors in the environment that affect health status, implements the plan of care, and evaluates the effects of that care.

Nurses are conscious that whatever they observe about their patients, and how the patients perceive what is happening to them, will be influenced by the setting in which it occurs and by the forces, obvious and subtle, that the setting contains. Nurses' education and experience have prepared them to subject all nursing phenomena to the analytic scrutiny that is part of the nursing process: nursing professionals look at individual trees as well as the woods. This component of excellent nursing practice goes beyond the mere counting, reducing to numerical scores, and checking off on lists of the results of observation (that Florence Nightingale identified as a nurse's most necessary skill), and implies a subjective determination. Rather than being whimsical or unreasoned, however, such a determination involves the combination of all parts into a meaningful whole in which theoretically and technically skilled and personally and socially perceptive health care practitioners engage as they make clinical judgments.[20]

As health care professionals scrutinize and analyze their data carefully, they examine the results of biological, medical, technical, and social advances that would have boggled the minds of their ancestors. This may shake, if not shatter completely, certain personal and professional beliefs that most of them value highly, such as convictions about the dignity of human life, the uniqueness of each human being, and the freedom of every individual to control her or his life and life style. As members of society and as practitioners, accountable health professionals may have (or be required) to rethink and redefine their purposes, their nature, and their value systems. To do this, they will require personal philosophies that are meaningful to them, explicit definitions of their ethical beliefs, identification of their own personal–professional conflicts, and acknowledgement of the extent to which they are imposing these on others and to which these conflicts affect the health care being provided.

Nurses constitute the largest group of health care professionals and hence they have the potential to influence very strongly the health care system, its practitioners, and its consumers. Nurses who are professionally accountable recognize that the direction of scientific discov-

---

[20]Edith P. Lewis, "Quantifying the Unquantifiable," *Nursing Outlook,* **24,** 147 (1976).

ery and its application to mankind are not out of their hands. They have personal, professional, and legal responsibilities to ask probing questions about scientific and technical research, and its application (or lack of it) in practice. Each health professional and each patient must be given the right, without censure or humiliation, to hear a drumbeat different from that of others. Professional practice must be directed by a genuine concern for a patient's own worth, for him or her as a person, regardless of that patient's social values and capacity for achievement.

This may seem to be a tall order, but is a professional imperative for health professionals, such as nurses, who allege that their hallmark is the caring quality of their behavior. In the process of caring, the center of attention is on the growth of another person, though during the process, the person caring actualizes her- or himself, and hence benefits along with the client.[21] The patient must be accepted as he or she is, and involved in the planning of care that is developed out of that patient's needs, health problems, the family and its resources, the character of health care system, and the resources of those health care workers who are available.

Nurses and other health professionals have knowledge and experience that suggest directions for care. At the same time, patients need to know why particular courses of action are being suggested, and need to have opportunities to accept, modify, or reject proposed plans of care without fear of being rejected by health care teams. As human beings, health workers are subject to bias that grows out of past experience, knowledge, perceptions of situations, feelings, values, and assumptions. By thinking critically about their work, health professionals can attempt to discipline the elements that affect their handling of data and thus enhance their ability to practice competently.

The position that the Government of Canada has taken on the responsibility of individuals for their health status, at a time when bookstores are replete with medical self-help literature and health care professionals are promoting autonomy and responsibility in patients, does not suggest a return to 1774 and to Dr. William Buchan's volume, *Domestic Medicine* (subtitled ". . . An attempt to render the Medical Art more generally useful by showing people what is in their own power both with respect to the Prevention and Cure of Diseases"), or to the Thomsonianism of the early part of the nineteenth century.[22] What it does, however, is challenge health professionals (and perhaps particu-

---

[21]Milton Mayeroff, *On Caring,* New York, Harper and Row, 1971, p. 30.

[22]Victoria Vespe Ozonoff and David Ozonoff, "On Helping Those who Help Themselves," *Hastings Center Report,* 7, 7–10 (February, 1977).

larly nurses, who have declared themselves to be *health* practitioners) to narrow Talcott Parsons' "competence gap" between the patient's and the health care worker's technical knowledge.[23] This could lead to a more humane and less exploitative health care system in which "patients' participation in therapeutic decision-making would be based on truly informed consent."[24]

The ethical imperative for the nurse who is accountable for professional practice is to ensure that there is provision for the exercise of action by patients that is based on their informed consent. Romanell has reminded nurses that the guiding principle for ethical behavior is epitomized in the words of Socrates in Plato's *Apology:* "The unexamined life is not worth living."[25] The examined life is lived by those nurses who are willing to question the prevailing customs and taboos of the situations in which they find themselves, including their own behavior, to identify whether what they see is consonant with the standards of practice for which they stand accountable.

In *The Present Crisis,* James Russell Loweli warned:

New occasions teach new duties: Time makes ancient good uncouth;
They must upward still, and onward, who would keep abreast of truth.

This challenges the accountable professional to subject established methods, policies, and institutions, including codes of ethics and tried and proven ways of acting, to constructive criticism in an effort to determine the need to transform the old order into a new and better one. Although this will not provide clear-cut mechanisms for solving all of the dilemmas of professional practice, it can stimulate the practitioner to strive for excellence in her or his own practice, to apply traditional ethical concepts that are adaptable to the culture or cultures in which one finds oneself, and thus to be able to make appropriate critical decisions when faced with ethical dilemmas.

---

[23]*ibid.*

[24]Howard Waitzkin and Barbara Waterman, *The Exploitation of Illness in Capitalist Society,* Indianapolis, Bobbs-Merrill, 1974, p. 116

[25]Patrick Romanell, "Ethics, Moral Conflicts and Choice," *Amer. J. Nursing* 17, 850 (1977).

24

Biomedical Developments and The Public Responsibility of Philosophy

David J. Roy

Center for Bioethics, Clinical Research Institute of Montreal, Montreal, Canada

I. THE ARGUMENT: AN OVERVIEW

The title of this presentation has been chosen to suggest that contemporary biomedical developments are generating a range of tasks which constitute a specific challenge to the community of philosophers. Theoretical and technical advances in the domain of biomedicine promise to increase enormously our power over the biological correlates of the human person and behavior and, indirectly, over the shape of those institutions that are central to the functioning of our communities and of human society in general. Moreover, the activities now permitted and also promised by the cumulative research underway in biomedical fields will, as is the case with science and technology generally today, set up chains of consequence and effect that touch all present human communities on this planet, and will reach into future generations.

The extent of this scientific–technological power has already stimulated explicit, if not highly systematized, reflection on the substance, scope, and limits of the scientific–technological community's public responsibility, i.e., its responsibility to the total human community. Great power, of course, demands effective norms to guide its use, to apportion its benefits justly, and to protect all who might be adversely affected by its risks. When the power in question promises to give a measure of rational and planned control over the design of human nature, over the kinds of human beings to be produced or

277

permitted to live, over the kinds of human behavior to be accepted in a society, then the norms needed must necessarily encompass judgments about what is humanly worthwhile and, more profoundly, about what is humanly essential. Norms to regulate responsibly a global activity capable of affecting the foundations and structure of human nature call for judgments about what is essentially or normatively human.

The identification of these norms and the attainment of the verified judgments essential to norms able to function as the basis for a global ethics represent tasks as urgent and important as the scientific and technological activities that give rise to the need for these norms. However, because the entire human community can expect to be affected, perhaps profoundly affected, by biomedical science and technology, it is in the name of this entire community that the elaboration of effective norms must be achieved. The clarification, identification, and grounding of ethical norms, particularly of norms so comprehensive in scope as to encompass judgments on what is essentially human, is a philosophical activity. The philosophical community is now challenged by the very course of research events to take up this work not only in its own name, not only in the name of academic interests and concerns, not only as an exercise in collegial collaboration with the scientific and technological communities, but, more comprehensively, as a moral response to the needs of the entire human community. This is what is meant by the public responsibility of philosophy vis-à-vis contemporary developments in biomedicine.

These reflections have defined the referent of the philosophical community's responsibility, those to whom the philosophical community is responsible, given the biomedical revolution and the questions this revolution is generating. But for what is the philosophical community supposedly responsible? What is it called upon to deliver?

The contributions expected of and required from the philosophical community should be as varied and mutually fructifying as are the specializations of method, analysis, and synthesis that constitute the highly differentiated and heterogeneous pattern of intellectual activity we call philosophy. There is not just one way of doing philosophy. There are many kinds of philosophical task, and we have historically and gradually developed philosophical methods to meet these tasks. Philosophy, of course, is itself in a process of evolution.

The focus here is somewhat restrictively on one of the philosophical tasks that appears to be most relevant to the mandate of public responsibility, a mandate to philosophy emerging from contemporary developments in science and technology, and, more specifically, from contemporary biomedical developments.

It is quite a widely accepted view that value judgments are doomed to be subjective and lack the objective character of scientific knowledge that assures the intersubjective validity of experimental results across political, cultural, and disciplinary frontiers and across interest and preference boundaries as well. What is purely subjective varies from subject to subject, from person to person, or at best from community to community. How, then, can one ever hope to discover a series of judgments able to function as the normative basis for a global ethics, able to provide a series of norms for biomedical science and technology that might possibly enjoy the consensus essential for a global ethics? The notion of science as the source of value-free, factual, objective, and intersubjectively valid knowledge would seem to rule out the possibility of discovering a rational foundation for a global ethics, unless, perhaps, a philosophical analysis could uncover the norm or set of norms that in fact functions as the basis of the entire scientific enterprise and for all scientific discourse worthy of the name.

A structure that cannot itself be proved but is the condition of the possibility of all proof, or which cannot be verified by the methods of science but is presupposed as the basis of all verification in science, represents a typical instance of what a transcendental analysis attempts to uncover. Such structures are called transcendental with respect to a given domain of human activity not because they are outside the reach of that activity but, on the contrary, precisely because they are implicated in every genuine instance of that activity. One of the tasks essential to philosophy's public responsibility consists precisely in carrying out the kind of transcendental analysis capable of uncovering a transcendental structure intrinsic to the scientific enterprise, and in determining whether and to what extent such a structure can function as the normative basis for a global ethics. The primary objective of this presentation is to indicate that such an analysis is possible and such a basis is available. It is not possible to carry out the whole analysis here. But even to indicate how the analysis works and what its results are should be sufficient to exemplify one of the pathways philosophy can follow to fulfill part of its responsibility to the human community at a moment in history marked by rapid and powerful advances in biomedical science and technology.

II. BIOMEDICINE, VALUES, AND THE
RECONSTRUCTION OF HUMAN NATURE

Medicine cannot be adequately represented simply as a neutral, fact-bound, value-free science operating within a restricted domain of nature or human nature. What counts as health, well-being, illness, disease, incapacitation, disability, handicap, or abnormality depends on a range of determinations of what is individually and societally worthwhile; it depends, in fact, on a range of very complex value judgments. Human values, individual and social, are an integral component of the medical enterprise.

It may well be the case that value judgments guide medicine silently, authoritatively, and efficiently most of the time. The trouble starts, and the need for a profound clarification of fundamental concepts, positions, and beliefs begins, when contradictory value judgments divide the medical and broader human communities on matters that are not marginal, but central, to human life and development.

In a range of instances, the power of biomedicine today, its ability to effect changes (with whatever value coefficients), reaches far beyond earlier, more restricted notions of therapy. The reconstruction of human beings as well as the indirect reshaping of institutions central to any human society—not simply the restoration of health and the containment of disease's ravages—increasingly appear to be less utopian, less science-fictional expectations of what contemporary biomedicine is and will be able to deliver.

Activities that touch upon this or that particular value of human well-being are one thing. Quite another, and an obviously much more urgent and difficult, matter is at stake when we confront sciences and technologies that critically raise the question of what "human" means and, even further, suggest answers very different from those we can identify across the long traditions of Eastern and Western cultures. To raise the question of the meaning of "human" is a challenge to form a judgment on a pattern of meaning and belief that traditionally functioned as the presupposed norm and basis of all judgment.

Contemporary advances across the many fields of biomedicine are revolutionary in character. These advances promise the power to effect change in the foundations of personal and societal being. Moreover, the mutual cross-fructification of many apparently unrelated levels of biomedical research and the cumulative power one can expect to emerge from this cross-fructification will inevitably shorten the response time available to the human community to prepare for and select from, rather than simply to adapt to, the changes and alternative courses for human development this power will render possible.

With radical alterations in the frameworks for human development come profound changes in our notions of human purpose. How do we set about ordering priorities and judging what is humanly desirable or imperative when the traditionally constant basis for these judgments, the structures of human personal and societal being, is now subject to designed change?

It would appear that the foundations for ethical norms, particularly the foundations linked to our normative images and concepts of what "human" means, are or will shortly be in a state of flux, subject to shifts—but in what direction? If the articulation and clarification of these foundations is an integral part of philosophical work, then the philosophical community faces new challenges and, indeed, a high and new mandate of public responsibility.

III. PARTICULAR ISSUES AS CHALLENGES TO PHILOSOPHICAL REFLECTION

Some of the developments in medicine today would *seem* to be philosophically prosaic: they *appear* simply to buttress and permit a more effective realization of traditional values.

An example. A short ten or fifteen years ago children afflicted with spina bifida and myelomeningocele stood little chance of surviving. Most died soon after birth, many quite rapidly. The same was true for babies along the whole spectrum of serious neonatal defects. Since the development of shunt techniques, refined surgical methods, powerful antibiotics, intensive care units, high specialization of medical competence, and technologically perfected life support systems, it is now possible to save—or at least prolong—the lives of these babies and of many other patients as well.

What modern medicine can do for these babies, however, so often seems far from being enough. Lives can be prolonged, but will the lives salvaged really be worth living? Are we, in all such cases, really dealing at least with human beings, if surely not with persons in the full sense of the term? Some judge that life for many of these babies will not be worth living, that they should not be treated and should be allowed to die. Some do not die, however, at least not quickly. May they be helped to die—quickly? Some would answer affirmatively because these babies in many cases cannot be judged to be human, at least according to certain suggested profiles of humanhood. So it is that one instance of apparently prosaic developments in medicine inevitably leads to questions of a high and difficult philosophical order.

This and numerous other examples lend support to the position that even "ordinary" medicine is part of a project that touches constantly upon the most profound and difficult of philosophical questions.

> Medicine is the most revolutionary of human technologies. It does not sculpt statues or paint paintings: it restructures man and man's life. Medicine changes the nature of man and like all revolutions, medicine has its own ideology. One might think it would be nice if it were otherwise, but every truly human endeavour requires a purpose, and a purpose involves a choice of values. . . . In short, medicine is not merely a science, not merely a technology. Medicine is a singular art which has as its object man himself. Medicine is the art of remaking man, not in the image of nature, but in his own image; medicine operates with an implicit idea of what man *should* be. Medicine is, consequently, value infected.[1]

A second example. Medical research and the medical project, even when their goals are relatively modest, often fall far short of success if we examine the extent to which those really in need are effectively reached. Though the treatments and medicine to control or cure leprosy are available, funds are unavailable to apply this treatment to the over ten million still untreated leprosy sufferers in the world. Is it reasonable to continue allocating enormous sums of money for basic research while we fail to deliver the cures it discovers across national boundaries to those in greatest need?

Why do we fail to achieve this delivery? Some find it difficult to clarify a strict moral obligation of any one nation to act on behalf of the afflicted in other nations. But do we really have a viable ethics of international relationships? Have we developed an ethics tailored to the concept of *world community,* a concept that surely is applicable today? We have not. And this deficiency owes in part to our lack of a philosophy of world community as a complement to our political philosophies tailored to the prides and prejudices of national interest. So it happens that a second example from a relatively routine domain of medical concerns takes us to the fringe of another kind of difficult and comprehensive philosophical question, namely, with respect to which human community do we measure our duties of justice and order our priorities in the distribution of our most necessary resources?

There are, of course, quite different developments in biomedicine

---

[1]H. Tristram Engelhardt, Jr., "The Philosophy of Medicine: A New Endeavor", *Texas Rep. Biol. Med.* **31,** 445 (1973).

that are or promise to be *philosophically dramatic,* i.e., promise to radically alter traditional institutions and patterns of behavior and relationship.

An example. Research and work already underway on the reproductive technologies deserve mention here as one set of instances among many others which merit consideration. In vitro fertilization techniques were initiated, in part, to devise a way to permit women who have healthy ovaries, but who are afflicted with other conditions preventing conception, to have their own babies. Work with this technique is very far advanced, indeed, up to the stage of embryo transfer. However, in vitro fertilization, if and when combined with the techniques of freezing embryos, embryo tissue cultures, methods of genetic analysis, and artificial placentae and amniotic tanks, could lead to a radical separation of human reproduction from sexual intercourse and its corresponding pattern of interpersonal relationship and the institution of marriage. The designed production of babies presumes many things, not the least of which is a desirable pattern or design.

The questions raised by these techniques—some already developed, others still in development—go far beyond the methods and horizons proper to any technology. A radical alteration in the origin of human beings raises questions about the nature of human beings. But short of this large question, even if closely related to it, are the equally pressing questions dealing with, for instance, the desirability or necessity of our institutions of marriage and family as the optimum matrix for the continuance of the human species.

A second example. Many highly educated people still find it difficult to suppress a smirk when listening to dedicated "death abolishers." Nevertheless, gerontology and its prolongevitists have already been taken seriously by many thinkers, and they may well have to be taken even more seriously in the not too distant future. Albert Rosenfeld's *Prolongevity* has convinced "a reviewer who doesn't particularly want to believe it, that some effective tampering with the aging mechanism is likely to be undertaken quite soon."[2]

We are already finding it difficult to do well by our older people. We are often at a loss as to how best to tap their pools of experience and to approach them with expectations that would honor their talents, their persons, and their dignity. Admittedly, we have not been very imaginative. Perhaps because of that, our economic resources and social institutions are even now strained by our attempts properly to care

---

[2]Robert E. Morison, "Tomorrow and tomorrow and tomorrow. . . . ," *Technol. Rev.* **79**(2), 75 (1977).

for the elderly. The strain may be expected to increase unless, of course, we can come up with some creative ideas.

Do attempts—and they may prove successful to a degree—to obtain a real extension of the normal life span qualify as one of these creative ideas? Approaching this notion or project with a simple risk–benefit calculus may do as a start. Soon, however, we move into questions calling for a sophisticated philosophical effort. How do we proceed to discuss the desirability, necessity, or purpose of a regular and frequent renewal of generations? Does death serve a function for the entire human community that we may suppress only at the cost of later and deep regret: Assume, for the moment, that we have slowed the aging mechanism significantly and confront massive problems as a result; could we then democratically decide to speed it up again? Would the majority of citizens be young or elderly?

Consideration of such scenarios may appear to be a waste of time to many. The research, however, is underway. We really do not know what response time is available to us to prepare. An even partially successful control or slowing down of the aging mechanism would entail profound changes in our social climate and the balance of our social institutions. At the extreme, perhaps even far short of that, "If we abolish death, we must abolish procreation as well, for the latter is life's answer to the former, and so we would have a world of old age with no youth, and of known individuals with no surprises of such that had never been before."[3]

This extreme point may well remain in the category of utopian fountain-of-youth dreams. However, the important and philosophically real issue here is that the possibility of even partial success in gerontological research

> ... raises questions that had never to be asked before in terms of practical choice, and ... no principle of former ethics, which took the human constants for granted, is competent to deal with them. And yet they must be dealt with ethically and by principle and not merely by the pressure of interest.[4]

And what should these principles be?

A third class of biomedical innovations, the realization of which is gradual and cumulative across many lines of research, and perhaps fairly imminent as a social phenomenon, may well prove to be *philo-*

---

[3]Hans Jonas, "Technology and Responsibility: Reflections on the New Tasks of Ethics," in *Philosophical Essays,* Englewood Cliffs, Prentice-Hall, 1974, p. 15.
[4]*ibid.,* p. 16.

sophically metadramatic. These innovations appear capable of taking us beyond, outside of, the human drama—capable of radically altering the nature or condition we have called "human" for thousands of years.

Techniques that intervene directly into central organ systems to modify or even design human behavioral responses might soon qualify for this category. However, the power to design new genetic combinations represents the single technological innovation that may have the longest-term consequences, bringing us to the frontier of pressing questions we will have no choice but to answer, but have never had to consider before. DNA recombination technology is young and immature, but this has not blocked a recent sharp and prolonged debate on whether these techniques should be employed at all, let alone developed further. The debate has, in fact, concentrated predominantly on the issue of biohazards, their magnitude, and likelihood of occurrence. To reduce both, guidelines have been proposed, specifically by the United States, Canada, and Great Britain.

Biohazards are obviously a central issue, an issue with philosophical implications. Another equally momentous issue is raised by the question of where the logic of DNA recombination technology will take us. May we reasonably expect that further refinements and developments of DNA recombination technology will soon give us the power to design or redesign human nature? Should we advance in this direction? If so, under what conditions? Are we prepared to assume control over the evolutionary process?

Some would laugh and doubt that so much is at stake. It seems infantile, however, to assume that DNA recombination techniques represent little more than some localized pattern of molecular technology, of prime interest to geneticists and molecular biologists, and with consequences limited to theoretical laboratory work and possibly a few significant agricultural and medical benefits. If this technology continues to develop as other technologies have developed—and the will of many to press in this direction is very strong—then it is reasonable to argue that

> . . . man will have a dramatically powerful means of changing the order of life. I know of no more elemental capability, even including the manipulation of nuclear forces. . . . It should not demean man to say that we may now be unable to manage successfully a capability for altering life itself.[5]

---

[5]Shaw Livermore quoted in William Bennett and Joel Guerin, "Science that Frightens Scientists: The Great Debate over DNA," *The Atlantic Monthly* 239(2), 59 (1977).

The development of DNA recombination techniques raises a range of issues that transcend the domain of those questions scientific method is equipped to answer. *One such question* deals with the possibility of our acquiring power over the evolutionary process. The point has been mentioned before but bears recall in the form of Robert Sinsheimer's series of sharp questions:

> How far will we want to develop genetic engineering? Do we want to assume the basic responsibility for life on this planet—to develop new living forms for our own purpose? Shall we take into our own hands our own future evolution? . . . Clearly the advent of genetic engineering, even merely in the microbial world, brings new responsibilities to accompany the new potentials. It is always thus when we introduce the element of human design. The distant, yet much discussed application of genetic engineering to mankind would place this equation at the center of all future human history. It would in the end make human design responsible for human nature. It is a responsibility to give pause, especially if one recognizes that the prerequisite for responsibility is the ability to forecast, to make reliable estimates of the consequence. Can we really forecast the consequence for mankind, for human society, of any major change in the human gene pool?[6]

A second question raised by DNA technology centers on the status of free, unfettered scientific inquiry. How are potentially conflicting interests to be balanced in a pluralistic society?

> Here, I think, we have come to recognize that there are limits to the practice of any human activity. To impose any limit upon freedom of inquiry is especially bitter for the scientist whose life is one of inquiry; but science has become too potent. It is no longer enough to wave the flag of Galileo.[7]

True, restrictions on the freedom to search for truth are both repugnant and dangerous. But is the genetic search only a search for truth, for knowledge? It is very difficult to segregate pure from applied knowledge, truth from applicable power. Great power calls for proportionate knowledge. But is knowledge, the result of specialized methods, enough? We arrive at a *third question,* deeply philosophical in character.

> We begin to see that the truth is not enough, that the truth is necessary but not sufficient, that scientific inquiry, the revealer of truth, needs to be coupled with wisdom if our object is to advance the human condition.

---

[6]Robert Sinsheimer, "Troubled Dawn for Genetic Engineering," *New Scientist* **68**, 150 (1975).
[7]*ibid.,* p. 150.

... In the nucleic acids of the cell we have penetrated to the core of life. When we are armed with such powers I think there are limits to the extent to which we can continue to rely upon the resistance of nature or of social institutions to protect us from our follies and our finite wisdom. Our thrusts of inquiry should not too far exceed our perception of their consequence. There are time constants and momenta in human affairs. We need to recognize that the great forces we now wield might—just might—drive us too swiftly toward some unseen chasm.[8]

The issues mentioned indicate that shifting value preferences and tricky value conflicts are perhaps the philosophically least significant phenomena cluttering the trail of these biomedical "advances." The prospect of modifying or "rationally" redesigning the very human structure of operations involved in all value preference and conflict constitutes, indeed, the more difficult and puzzling challenge to philosophy. Is it philosophically possible to unfold a normative ethics capable of differentiating between those projects and changes that may be expected to enhance and those projects and changes that carry great risk of dismantling the structures we refer to as "the human condition"?

IV. THE FOUNDATION OF ETHICS AS THE PUBLIC RESPONSIBILITY OF PHILOSOPHY

I am indebted to Hans Jonas and Karl-Otto Apel for the notion that philosophy indeed has a public responsibility. How this responsibility is to be conceived and exercised is bound to vary with one's notion of what it is that both professional and the broader human communities may reasonably expect from the community of philosophers. What response can one expect or demand from philosophy? Indeed, what is the question that is seeking a philosophical response?

V. AN ETHICS GROUNDED IN THE NATURE OF THINGS

Hans Jonas has mentioned his being recalled from "theoretical detachment to public responsibility" and his finding a new task for his philosophizing. The source of the call was a question latent in the "growing realization of the inherent dangers of technology as such—not of its

---

[8]From a talk given by Robert Sinsheimer at the University of California, June, 1976 and quoted in Nicholas Wade, "Recombinant DNA: A Critic Questions the Right to Free Inquiry," *Science* **194,** 304 (1976).

sudden but of its slow perils, not of its short-term but of its long-term threats, not of its malevolent abuses that, with some watchfulness, one can hope to control, but of its most benevolent and legitimate uses which are the very stuff of its active possession."[9] The latent question is really a quest for an effective ethics for a technological culture marked and shaped by fundamental changes in the characteristics of human action. "Modern technology has introduced actions of such novel scale, objects, and consequences that the framework of former ethics can no longer contain them."[10]

How can one best pursue this quest? An effective ethics would have to be based upon and emerge from a philosophy of organism, a philosophy of mind, and, more generally, a philosophy of nature. In a word, the "ontological quest" pursued in the direction of human action, technically potentiated, as "reality" becomes a quest for an effective ethics.

> . . . Only an ethics which is grounded in the breadth of being, nor merely in the singularity or oddness of man, can have significance in the scheme of things. It has it, if man has such significance; and whether he has it we must learn from an interpretation of reality as a whole, at least from an interpretation of life as a whole. But even without any such claim of transhuman significance for human conduct, an ethics no longer founded on divine authority must be founded on a principle discoverable in the nature of things, lest it fall victim to subjectivism or other forms of relativity. However far, therefore, the ontological quest may have carried us outside man, into the general theory of being and of life, it did not really move away from ethics, but searched for its possible foundation.[11]

The search for the foundations of ethics is a philosophical task and responsibility, and demands—so this view would hold—"the continuation of the ontological argument into ethics".[12] The course of science and technology across the biological revolution has transformed this philosophical task into an "ethical duty."[13] The responsibility of philosophy, then, would demand its effective grounding of ethics in the nature of things, including, so it would seem, the nature of man.

But how can philosophy hope to discover these foundations for an ethics capable of guiding our use and development of science and

---

[9]Hans Jonas, "Introduction", in *Philosophical Essays,* p. xvi.

[10]Hans Jonas, "Technology and Responsibility: Reflections on the New Tasks of Ethics," *op. cit.,* p. 8.

[11]Hans Jonas, "Epilogue", in *The Phenomenon of Life,* New York, Harper and Row, 1966, p. 284.

[12]Jonas, "Introduction", *op. cit.,* p. xvii.

[13]*ibid.,* p. xvii.

technology if the very sources of the principles for such a foundation —the nature of man and the nature of things—are subjected to radical change by the very activities we seek to control? Hans Jonas would phrase the dilemma even more sharply in terms of the survival of candidates for a moral order.

> The *presence of man in the world* had been a first and unquestionable given, from which all idea of obligation in human conduct started out. Now it has itself become an *object* of obligation—the obligation namely to ensure the very premise of all obligation, i.e., the *foothold* for a moral universe in the physical world—the existence of mere candidates for a moral order.[14]

VI. A CRITICAL QUESTION

If the presence of humankind in the world has become the object of moral obligation and no longer functions as the unquestionable basis of the moral order, and further, if human nature can now or soon be subjected to "rationally planned" and designed change as opposed to its previous status as an invariant natural norm for human activity, how do we set about delivering a viable and solid foundation for ethics in a scientific–technological culture that is rapidly gaining global power as well as power over future generations? This question seeks a grounded and intersubjectively-valid global ethics. Nothing short of this is needed now to match the reach of our scientific and technological power. Can philosophy deliver the foundations for such an ethics? We should not answer too rapidly. Jonas, for one, has felt constrained to admit: "Philosophy, it must be confessed, is sadly unprepared for this —its first cosmic—task."[15]

VII. TRANSCENDENTAL ANALYSIS, THE A PRIORI OF COMMUNICATION AND THE FOUNDATIONS OF A GLOBAL ETHICS

Why is it that philosophy is so sadly prepared for its cosmic task? And more precisely, what is the cosmic task?

---

[14]Hans Jonas, "Technology and Responsibility: Reflections on the New Tasks of Ethics," *op. cit.,* p. 12.

[15]Hans Jonas, "Seventeenth Century and After: The Meaning of the Scientific and Technological Revolution", in *Philosophical Essays,* p. 80.

Karl-Otto Apel has suggested an answer to both questions with his reflection on the central paradox plaguing our search for an effective, i.e., intersubjectively valid, ethics in an era characterized by scientific and technological activity having global and long-term consequences.[16]

VIII. THE PARADOX

Scientific and technological activity affects the vital interests of nearly everyone on this planet today and is already setting up chains of consequences which will affect future generations. Responsibility demands an adequate accounting in advance of how we will affect others by what we are doing, particularly when life and the conditions for life and human development are at stake. So we need a global ethics as we never have before.

Arriving at an ethics, at a series of norms or regulative principles, that might enjoy consensus across national and cultural boundaries, however, seems to be nearly impossible. Rational justification of norms would have to be at least a necessary, if not sufficient, condition for transnational and transcultural consensus.

The notion of value *judgments* as opposed to subjective and arbitrary value *preferences* is foreign to the currently dominant notion of scientific objectivity, though. A rational, i.e., objective, i.e., intersubjectively-valid, justification of norms and morally regulative principles would seem to be in principle unattainable; so also a global ethics. Thus runs a general description of the paradox.

IX. THE OPENING QUESTION

Our need for a global ethics to manage responsibly a network of scientific and technological activity that, particularly with the biomedical revolution, reaches potentially into every ecological niche and into every human life defines an essential dimension of philosophy's cosmic task: to deliver the foundations for an ethics capable of matching the macro-reach of our activity and its consequences. Philosophy is sadly unprepared for this task, primarily because it has largely abandoned foundational analysis in favor of a value-free metaethics, one that gener-

---

[16]Karl-Otto Apel, "Das Apriori der Kommunikations-Gemeinschaft und die Grundlagen der Ethik," in *Transformation der Philosophie,* Band II, Frankfurt am Main, Suhrkamp Verlag, 1973, pp. 359ff.

ally operates on the scientistic assumption that transculturally valid rational argument is restricted to the notion of objectivity characterizing the work of the formal sciences of mathematics and logic as well as the operations of the empirical and analytical "hard" sciences. On this assumption, norms are attained by subjective and arbitrary decision or choice, not by rational argument. In a word, the extreme conclusion of this assumption is that ethical norms are not possible, let alone a global ethics.

X. THE TRANSFORMATION OF PHILOSOPHY

How, then, is it possible for philosophy to take up and successfully pursue the foundational task that defines a central dimension of its public responsibility today? An adequate sketch of how this is possible and of how it is to be done would demand a tracing of the transformation philosophy has undergone and is still undergoing. That tracing would have to identify

> . . . the dialectical contacts which in fact constitute a pattern of relationship between the *Linguistic Tradition* (linguistic analysis, the logic of language, logical atomism and logical positivism, constructive semantics, pragmatic semiotics—Russell, Wittgenstein, Carnap, Pierce, Morris), *The Hermeneutical Tradition* (Schleiermucher, Ditthey, Heidegger, Gadamer) and *The Tradition of Transcendental Analysis* (Kant across variations of the method up to Lonergan).[17]

These dialectical points of contact and the pattern of relationship they constitute also sketch the development philosophy is undergoing toward *dialog as method* to integrate the varied functions of philosophical reflection and to relate effectively the heuristic structures that deliver our understanding of the universe.[18] The historical unfolding of the linguistic tradition across a threefold transformation of the criterion of meaning—from an empirical to a logical to a pragmatic criterion of meaning—has taken even this one tradition a long way toward a recognition of dialog as method, and toward a recognition of the complementarity of hermeneutical and transcendental method.

Obviously, even sketching this transformation of philosophy

---

[17]David J. Roy, "Is 'Philosophy' Really Possible? A Meditation on Heidegger and Wittgenstein with Karl-Otto Apel", *Laurentian Univ. Rev.* **9**, 82 (1977).

[18]Cf. the entire issue of *Neue Hefte für Philosophie,* 2/3 (1972), edited by Rüdiger Bubner, Konrad Cramer, and Reiner Wiehl and devoted to the theme, "Dialog als Methode."

would take us beyond the scope of this presentation. Enough has been indicated to permit the summary of a more direct approach toward the foundation for an intersubjectively-valid global ethics.

XI. TRANSCENDENTAL ANALYSIS AND THE MORAL NORMS PRESUPPOSED BY SCIENTIFIC DISCOURSE

Working toward the rational foundations for a global ethics demands parting ways with the assumption that "scientific rationality, in the sense of the logic of science (i.e., logical inferences plus observation), in fact exhausts the whole of human rationality, so that beyond its limits only the irrationality of arbitrary decision can exist."[19]

How philosophy can part ways with this assumption and move towards the foundations of ethics may be summarized in the following steps. At this point, I follow Karl-Otto Apel's analysis very closely.[20]

1. Though an acceptance of the normative character of logic and of the logic of science does not entail the view that logic implies an ethics, it is difficult to argue convincingly against the proposition that logic—and along with logic, science, and scientific technology—*presupposes* an ethics. The logical validity of argumentation cannot be verified without in principle presupposing a community of thinkers capable of intersubjective understanding and consensus. A community of argumentation is the necessary real and pragmatic presupposition for the verification of *any* argument. The success of any stretch of argumentation, then, demands not only that one follow the *rules of logic,* but presupposes a real, if not always completely explicit, acceptance of a rule of behavior, a moral rule: lying as well as a dogmatic refusal to unfold and justify the argument for a proposition or claim would destroy the dialog that is constitutive of the community of argumentation presupposed for the verification of the validity of one's argument.

Turning to logic for the construction of any proposition, claim, or set of propositions and claims encompasses an intention to seek and achieve unbounded verification of one's argument. This demands the recognition in principle of all members of the community of argumentation as dialog partners with equal claim to involvement in the argumentation process. This demand becomes more tangible if the proposition

[19]Karl-Otto Apel, "The A Priori of Communication and The Foundation of the Humanities," *Man and World,* 5 7 (1972).

[20]Karl-Otto Apel, "Das Apriori der Kommunikations-Gemeinschaft und die Grundlagen der Ethik", *op. cit.,* pp. 395ff.

or claim is not purely theoretical or verbal in character, but takes an active or technological form touching not only upon the ideas but also the interests, perhaps vital interests, of members of this community.

2. The characteristic of a transcendental analysis is that it attempts to uncover the necessary conditions for the possibility of any given instance of activity. The fact that logic is employed to uncover a condition of the possibility for logical argumentation does not mean that the foundational analysis employed in this "uncovering" has to function as a process of deduction within the framework of an axiomatic system, much less within the framework of the same axiomatic system we would presumably be attempting to ground. Structures that cannot themselves be proved because they are the condition of the possibility of all proof are precisely the objects of a transcendental analysis.

3. If logic presupposes an ethics, so also does science. Science demands a definite commitment to a community of experimentation, a commitment that necessarily carries coefficients of self-surrender, which may be summarized in the notion of a disinterested pursuit of knowledge and truth. The ethics of science demand a certain subordination, if not surrender, of personal, individual interests to the scientific-interest characteristic of the scientific community. Moreover, science employs logic and, as discourse, presupposes the ethics peculiar to logic, the ethics that permit a community of argumentation to function.

4. However, a closer examination should reveal that the community of argumentation presupposed by logic and by scientific discourse is not simply a community of scientists and logicians. "Science" stands not simply for a set of methods, specializations, and propositions. The scientific interest is itself only one among many characteristic of the human community. Over and above particular scientific propositions, the scientific interest itself and its claims must be argued and justified. The community of argumentation, presupposed by logical argument and scientific discourse, demands not only an attempt to achieve unbounded verification of propositions but, equally so, an attempt to achieve unbounded justification of interest claims.

The process of rational argumentation as the counterpoint to a dogmatic or obscurantist imposition of propositions, views, beliefs, and claims reaches very far. Its demands are high. The process demands at least an implicit recognition of all possible claims of all members of the community of communication that can be justified by reasonable argument. It demands, secondly, that one be prepared to justify all one's own claims by reasonable argument, indeed, before the entire community of communication. This community of communication and the ethic of recognition that is intrinsic to and constitutive of this community would seem to be a fundamental condition of the possibility of rational argumentation and scientific discourse.

XII. TWO PRINCIPLES AS A FOUNDATION FOR A GLOBAL ETHICS

The community of communication with its constitutive ethic that a transcendental analysis attempts to uncover as the condition of the possibility of rational argumentation and scientific discourse is, it must be emphasized, historical in character. The real community within which all discourse is embedded is all too often marked by the ego and group bias that block the quest for unbounded verification and justification of propositions and claims.[21] Claims can hardly be argued if they are systematically misunderstood. And enduring bias generates systematic misunderstanding. So rational argumentation and scientific discourse presuppose, and indeed contribute to, the realization of an ideal community of communication in which the limits to understanding, verification, and justification of propositions and claims are increasingly transcended.

If a community of communication with its constitutive ethic is accepted as the condition of the possibility of all rational discourse, then two principles would seem to follow as central to the strategy of a global ethics.

Principle 1. Every activity of individuals or communities should be directed to the survival of the human species as the real community of communication.

Principle 2. Every activity of individuals or communities should contribute to the increased realization of the ideal within the real community of communication.

These two principles are linked: the first is the condition of the possibility of the second, and the second defines a meaningful purpose for the first.

In a word, *the community of communication as the presupposition of all rational discourse constitutively demands an ethic of survival for the sake of liberation.* This of necessity must be a global ethic. It is an

---

[21]B. Lonergan, *Insight: A Study of Human Understanding,* New York: Philosophical Library, 1957, pp. 218–238.

intersubjectively-valid ethic wherever rational discourse is accepted as necessary.[22]

XIII. CONCLUSION

A summary of how philosophy can approach one of the tasks essential to its public responsibility does not cover all of the tasks, much less the full scope, of that responsibility. Working out the foundations for a global ethics is not the same thing as working out the framework of a global ethics. This second task calls for what Hans Jonas has identified as continuing the ontological argument into ethics. The ontological argument, of course, heads toward an integrated view of the universe. If this is so, it is essential to realize that "philosophy obtains its integrated view of a single universe, not by determining the contents that fill heuristic structures, but by relating the heuristic structures to one another."[23]

The sciences, of course, are instances of heuristic structures, and so also are all systematized forms of human discourse. Relating these structures, balancing them, and integrating them involves more than philosophy as architecture. For, to each heuristic structure corresponds a pattern of interests anchored in very real seekers. Relating heuristic structures without relating the seekers is to do only half the philosophical task. The full measure of philosophy's public responsibility calls for an effective realization of dialog as method. But can the community of philosophers carry the burden of that responsibility alone?

---

[22]The principles mentioned above come directly from Karl-Otto Apel, "Das Apriori der Kommunikations-Gemeinschaft und die Grundlagen der Ethik", *op. cit.*, pp. 431–432.

[23]Lonergan, *op. cit,.* p. 426.

Subject Index

297